routers

Elu

good
wood
routers
ALBERT JACKSON AND DAVID DAY

Collins

GOOD WOOD ROUTERS
Conceived, edited and designed at Inklink, Greenwich, London

Text: Albert Jackson and David Day

Design and art direction: Simon Jennings

Technical contributor and consultant: John Perkins

Text editors: Ian Kearey and Albert Jackson

Illustrators: Robin Harris and David Day

Studio photography: Paul Chave, Ben Jennings and Neil Waving

Indexer: Ian Kearey

First published in 1996
by HarperCollins Publishers, London

This paperback edition first published in 2002
by HarperCollins Publishers, London

Visit our website at www.collins.co.uk

A CIP record is available from the British Library

ISBN 0 00 713978 0

Text set in Franklin Gothic Extra Condensed, Univers Condensed and Garamond Book Condensed by Inklink, London

Printed in Singapore

CONTENTS

INTRODUCTION

Most power tools are no more than mechanized versions of their hand-held predecessors. They take much of the graft out of woodworking, but the business end of these tools remains much the same – a saw cut is still a saw cut, a hole is just a hole, and a screw will do its job well, whether it is inserted by hand or with a power screwdriver. The power router, on the other hand, is in a different category. This is a tool that has revolutionized the home workshop. It has all but replaced the specialized hand planes used for moulding, rabbeting and grooving, and works with a degree of accuracy that was once only the prerogative of the most skilled craftsperson. Combine all these functions and you can cut a range of woodworking joints, follow templates, trim plastic laminates, even carve wood sculpture. The remarkable power router is nothing less than a miniature machine shop.

ACKNOWLEDGEMENTS

The authors would like to thank the following for the supply of reference material and equipment used in this book.

Jim Pankhania,
Elu Power Tools Ltd,
Slough,
Berks

Steve Phillips,
Trend Machinery and Cutting Tools,
Watford,
Herts

John Roberts,
Robert Bosch Ltd,
Denham, Uxbridge,
Middx

Kenneth Grisley,
Leigh Industries Ltd,
Port Coquitlam,
BC, Canada

Ramon Weston,
Leigh Industries (UK) Ltd,
Chippenham,
Wilts

Martin Godfrey,
Woodrat,
Godney, Wells,
Somerset

John Perkins,
Wool,
Dorset

Technical contributor and consultant

John Perkins is a freelance writer for practical magazines, including *Routing*, which he helped set up, *Practical Woodworking* and *Practical Householder* (as Project Editor). He has also taught production techniques and workshop practice, and worked as a boat and yacht joiner. He works closely with Trend Machinery and Cutting Tools and with Elu Power Tools, as a writer and consultant on routers and power woodworking, and runs a design practice in Dorset, specializing in architectural and interior work.

Photography

The studio photographs for this book were taken by Paul Chave, *with the following exceptions:*

Ben Jennings, pages 100 (T and B), 105 (TR), 107 (TR

Neil Waving, pages 10 (TR), 88, 91, 95, 102 (TR), 106 (TR), 108 (TR)

The authors also acknowledge additional photography by, and the use of photographs from, the following individuals and companies.

John Perkins, pages 86 (C), 97(C), 110 (TR), 112 (TR)

Trend Machinery and Cutting Tools Ltd, pages 44 (TL), 66 (C & TR), 70 (BC), 71 (TL), 80 (TR)

Key to credits
T = top, B = bottom, L = left, R = right, TL = top left, TC = top centre, TR = top right, CL = centre left, C = centre, CR = centre right, BL = bottom left, BC = bottom centre, BR = bottom right

Jigs and guides

All shop-made jigs and guides designed by John Perkins.

CHAPTER 1 Unlike stationary woodworking machinery fitted with powerful motors, the lightweight portable router employs speed rather than power to produce clean, accurate cuts in wood, plastics and soft metals. A high-speed electric motor drives a precision-ground cutter that is used with a range of innovative fences, guides, templates and jigs, some of which are supplied with the router.

POWER ROUTERS

There are two basic types of power router – plunge and fixed-base. On fixed-base routers, the motor housing, which is clamped to the base, slides up or down to adjust the depth of cut, and the cutter projects permanently from the base. While this presents few problems when cutting work from an edge, starting and finishing a cut such as a stopped housing is more difficult for the less-experienced craftsperson, since the entire router must be lowered onto the wood and lifted clear while the cutter is still rotating.

For ease of use, woodworkers new to routing would probably do best to purchase a good-quality, general-purpose plunge router. Here, the motor rises and falls on a pair of columns rigidly mounted onto the base plate. Return springs counter the weight of the motor, allowing you to plunge the cutter square into the surface of the work and to retract it safely above the base plate before lifting the router from the work.

Fixed-base router

POWER RATINGS

Routers are rated by their power input; those used for light machining operations start at around 600W (¾HP), increasing to the 1850W (2½HP) machines used mainly in heavy industry.

However, power is not always the prime consideration when purchasing a router. A heavy-duty router will be relatively weighty and cumbersome, and may be difficult to handle, even for experienced machinists. In addition, if you want to use small, delicate cutters, remember that these are vulnerable to damage under too-heavy power loads.

Note also that horsepower is not as accurate a measurement of power input as wattage.

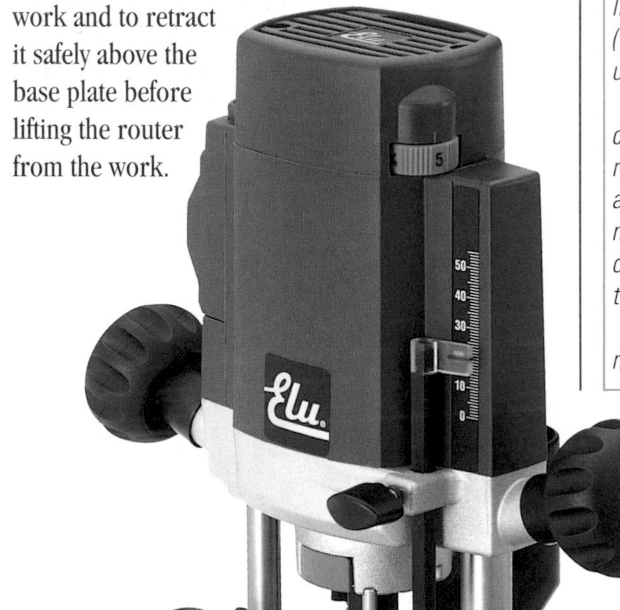

Plunge router

Demountable routers
On many light-duty plunge routers, the motor housing can be separated from the base, allowing you to mount the motor in an overhead stand or drill press. When using these routers by hand, ensure that the two parts are securely locked together, as the downward plunge action can dislodge the motor housing, forcing it up and out of the base unit.

ROUTER COLLETS AND SHANK SIZE

To hold the cutter, the router spindle is fitted with a simple chuck that takes a specific-diameter cutter shank. A locknut is tightened onto the threaded spindle, pressing a tapered and slit collet or cone into an internal taper ground in the spindle end. This closes the collet to grip the cutter shank firmly.

Light- and medium-duty routers use 6mm (¼in) shank cutters. Major manufacturers offer a range of interchangeable collets to allow for minor variations in shank size, but any specific shank diameter must only be used with a collet of the matching diameter.

Many medium-duty routers can also be fitted with 8mm (⅜in) collets and correspondingly larger and more-complex cutters. Before using such a cutter, check the manufacturer's recommendations, and make sure that they can be used in hand-held operations.

Heavy-duty routers mainly use 12mm (½in) shank cutters, but can also be fitted with smaller-diameter collets. When using small-shank cutters in these routers, ensure that the rate of feed into the workpiece and the load exerted on the cutter are not excessive, or the cutter may shear from its shank.

Collets are precision-ground to hold the cutter shank in perfect alignment along the motor-spindle axis. Any eccentric cutter rotation creates vibration in the router, resulting in a poor finish and possible damage to the cutter and router. The multi-slit collets supplied with good-quality routers align the cutter more accurately and grip the shank more securely than the split-cone collets found on less-expensive machines *(see page 28).*

Medium-duty routers
The medium-duty range starts at 750W (1HP); a good-quality plunge router of around 800 to 1200W (1¼ to 1¾HP) is ideal for furniture making, wood machining and some joinery work. These lightweight, compact machines are also suitable for inverted or overhead-table routing, and many accessories, jigs and templates are now produced for this size of router.

Heavy-duty routers
With power-input ratings of up to 1850W (2½HP), heavy-duty routers use strong, large-diameter-shank cutters for construction work such as window and door manufacture, and general joinery. There is little risk of overloading the motor, and you can make relatively deep, wide cuts.

Light-duty routers (above)
Inexpensive, low-power (up to 750W or 1HP) light-duty plunge routers can carry out most operations, including grooving, rabbeting and decorative moulding. However, buy only a good-quality router or you may find that the amount of time and effort required for even basic operations may be frustrating. The range of cutters available for light-duty routers may be limited.

Trimming routers (right)
Some light-duty routers are manufactured solely for trimming the edges of veneers and plastic laminates. Since they are made to work with a high degree of precision, they tend to be expensive and are not suitable for general routing.

ROUTER CONTROLS

All routers are fitted with similar controls, although there may be some variation in their positioning and operation.

Handgrips
These generally consist of two side handles for gripping and guiding the router. Some light-duty laminate-trimming routers may be fitted with a single handle. For good control and balance, the handgrips should be positioned as near to the base of the router as possible.

On/off switch
This switch should be located near one of the handgrips, so that you can reach it without having to release your control over the router.

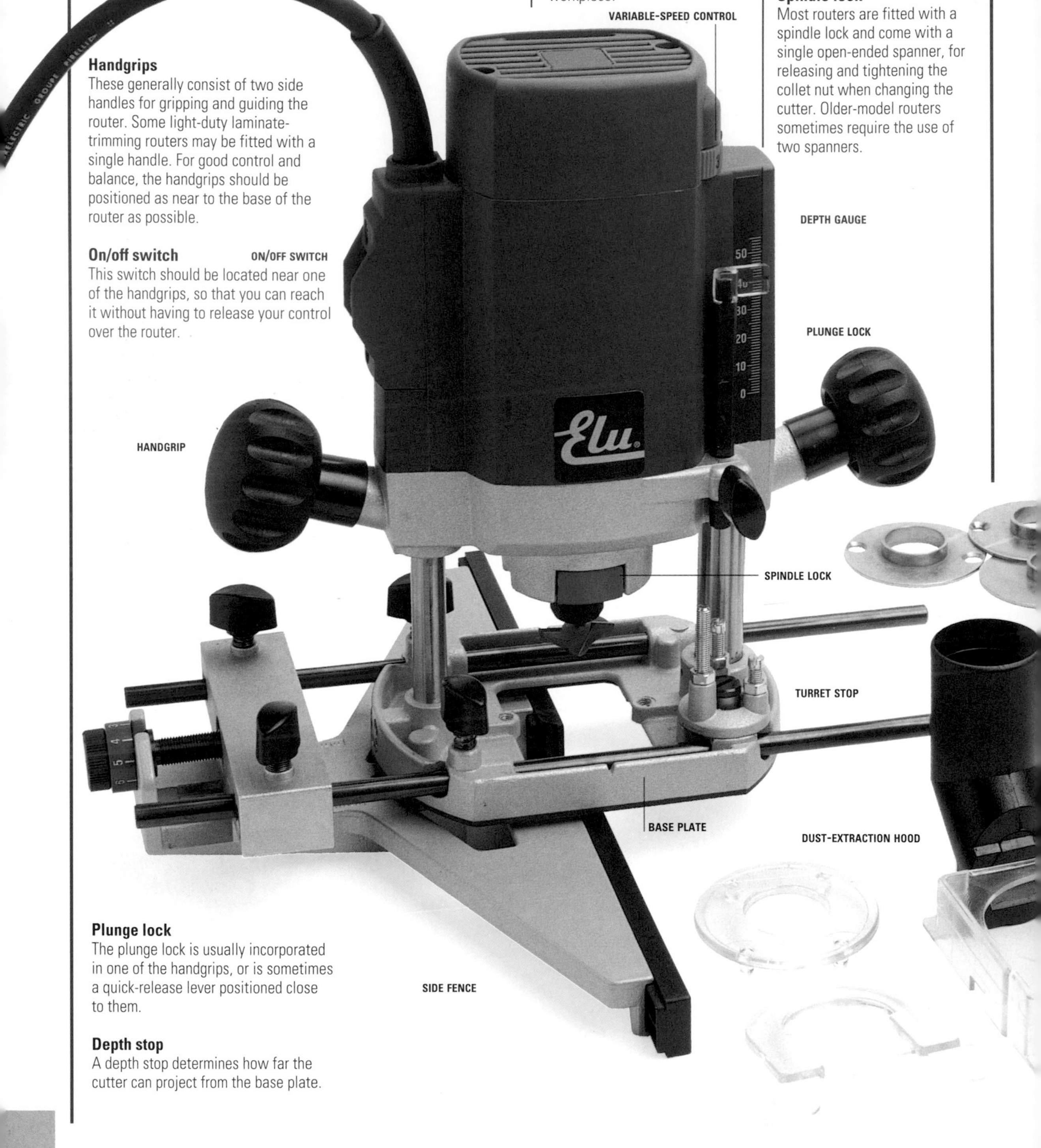

Plunge lock
The plunge lock is usually incorporated in one of the handgrips, or is sometimes a quick-release lever positioned close to them.

Depth stop
A depth stop determines how far the cutter can project from the base plate.

Turret stop
When routing in stages, you can pre-set the turret stop to make quick alterations to the depth of cut.

Side fence
Routers are supplied with a guide fence for edge-trimming and moulding, and for making cuts parallel to the edge of the workpiece.

Variable-speed control
This control usually consists of a simple dial showing five or six positions relating to specific speeds (the actual speeds differ between makes and models). Check that the variable-speed control is not positioned where it could be moved accidentally while routing.

Spindle lock
Most routers are fitted with a spindle lock and come with a single open-ended spanner, for releasing and tightening the collet nut when changing the cutter. Older-model routers sometimes require the use of two spanners.

VARIABLE-SPEED CONTROL
ON/OFF SWITCH
DEPTH GAUGE
PLUNGE LOCK
HANDGRIP
SPINDLE LOCK
TURRET STOP
BASE PLATE
DUST-EXTRACTION HOOD
SIDE FENCE

SPINDLE SPEEDS

Maximum router-spindle speeds are between 20,000 and 30,000 rpm, depending on the router's power output. Many routers are available with an electronic variable-speed control (see opposite), which automatically monitors and maintains the selected cutting speed, even when the feed rate changes. Because variable-speed routers run relatively quietly and smoothly, they are more comfortable to work with than single-speed models, and the 'soft-start' feature minimizes the initial jolt when starting the high-speed motor.

Although you can use the maximum speed for most operations requiring small-diameter cutters, some hard or troublesome timbers may call for a little experimentation to select the best speed for the job.

Large-diameter cutters have a high peripheral speed, and should therefore be run at lower spindle velocities. Comparatively low speeds should also be selected when cutting soft metals and plastic materials, to avoid overheating, weld-back and fusion of waste particles to the cutter (see page 34).

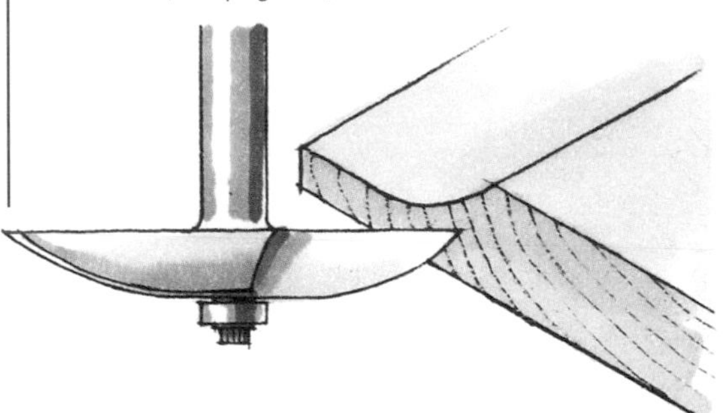

Base plate
In many routing operations, the router is guided by running the base plate against a straightedge, template or jig. Base plates are either completely circular and concentric to the spindle axis, or have at least one straight edge. The plate also carries cramps to secure the side-fence mounting rods, and threaded mountings to take other guides and accessories. Check that the underside of the base plate is fitted with a thin, replaceable plastic facing to avoid marking the face of the workpiece.

Standard accessories
Most routers are supplied with one side fence (see opposite), and a single guide bush, for use with templates and jigs. You can also buy many accessories, including curved-edge and circle-cutting guides, a riser plate for lifting the router above the work surface when trimming applied edge lippings and laminates, and an edge follower for edge-moulding and trimming curved or irregular-shaped work. Manufacturers often tempt the purchaser by including some accessories as part of the initial package.

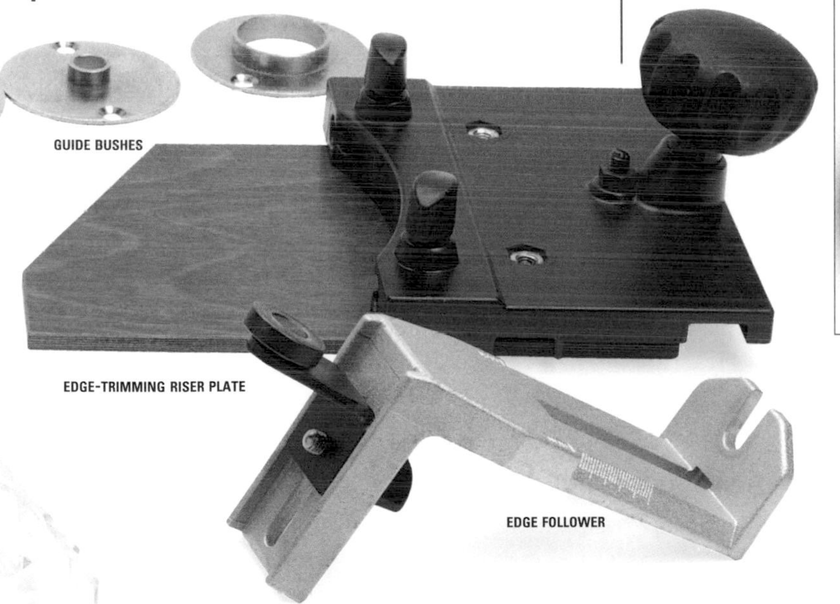

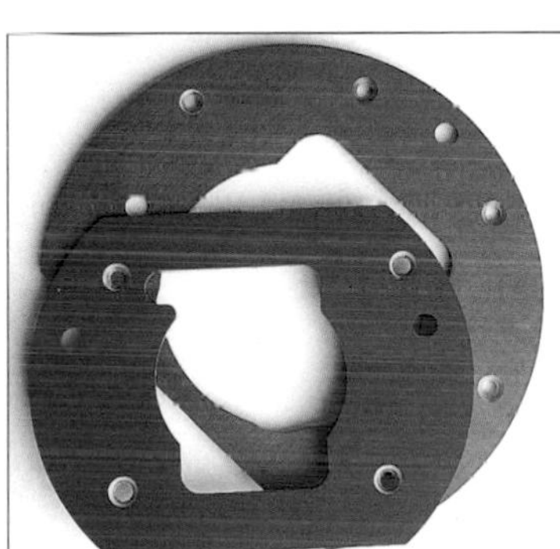

Base-plate facings

Dust extraction
Dust-extraction facilities are now supplied with all routers. They usually consist of a clear-plastic hood fitted on the base of the router, enclosing the cutter. The hood is connected by a flexible hose to a vacuum extraction unit, which collects waste material as soon as it leaves the cutter (see page 124). Extractors have been omitted from most illustrations, for clarity.

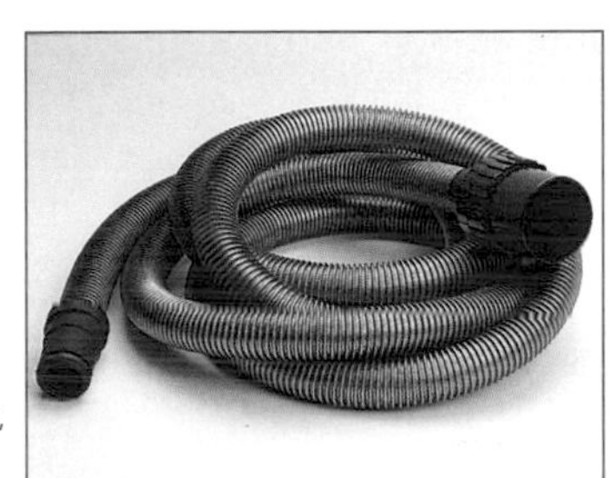

Dust-extraction hose

MAINTAINING ROUTERS

Because it has a direct drive, the router is a fairly simple machine to maintain. However, avoid dismantling it unless this is absolutely necessary.

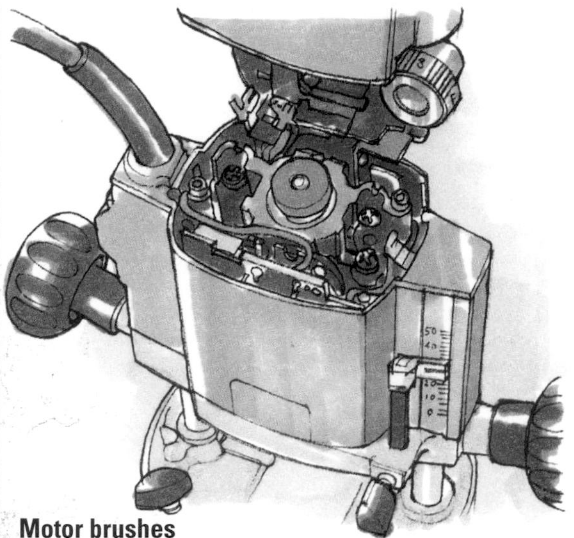

Motor brushes

Check motor brushes regularly, although any deterioration will be indicated by excessive sparking inside the top section of the motor housing. Replacing worn brushes is a fairly easy job, requiring the removal of either the external brush caps or the top section of the motor housing. If further dismantling is required, this is best left to a service engineer.

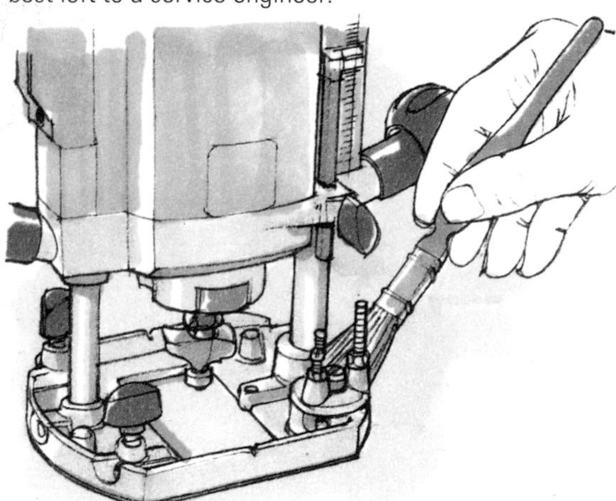

Removing dust

Use a vacuum cleaner to remove wood dust, after loosening it from tight corners with a small, stiff paintbrush. Attempting to blow out dust with a compressed-air line may force it further into the machine bearings. Remove resin deposits with a solvent such as white or mineral spirits, but keep solvents away from the motor, motor windings and switch.

Bearings

The bearings are sealed for life and should not require lubrication, which may have the effect of thinning the pre-packed grease. Replacing bearings is again best left to an experienced service engineer, who will have the correct tools and equipment for reseating and aligning them. If you are concerned about the bearings (for instance, a change in sound when running, slight vibrations, or play in the spindle), have them changed immediately, to avoid further wear on the spindle and damage to the armature and brushes.

Switch

Make sure the switch operates positively. Always have a faulty switch replaced, to minimize any risk while the router is in use.

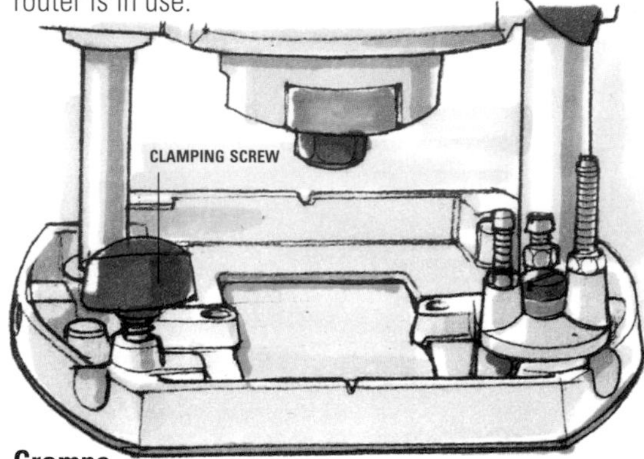

Cramps

Check that the clamping screws for side-fence rods are fitted with anti-vibration springs, to prevent them working loose.

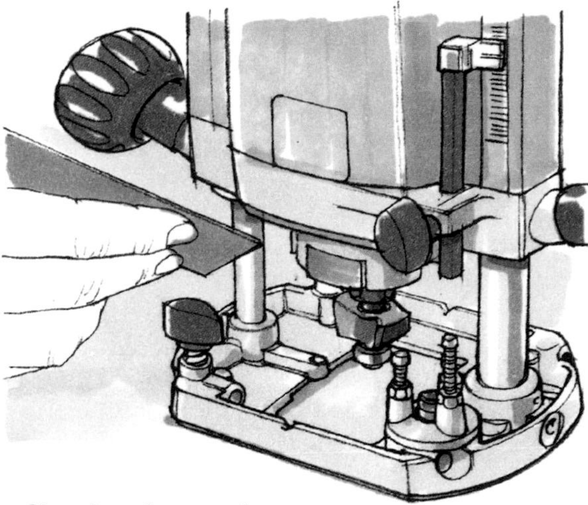

Cleaning plunge columns

Regularly remove dust and deposits from the plunge columns, and apply a light oil, silicone- or dry-lubricating (PTFE) spray. Scrape off deposits with the edge of a piece of plastic or stiff card. Do not use wire wool or abrasive paper on the columns, as these leave scratches to which resin and dust can adhere.

CHAPTER 2 Using the extensive range of cutters, or bits, available for the router, you can perform many moulding, rabbeting, housing-cutting and jointing operations. Router cutters are expensive items, especially the larger and more complex profiles, so it is worth buying good-quality cutters that will work well for a long time.

THE BASIC CUTTER

Router cutters should not only cut cleanly, but also clear waste material freely and quickly. If they do not, waste packs around the cutter and creates excessive friction, causing the cutting edges to overheat. The result is a loss of tempering in the steel and scorch marks on the surface of the work.

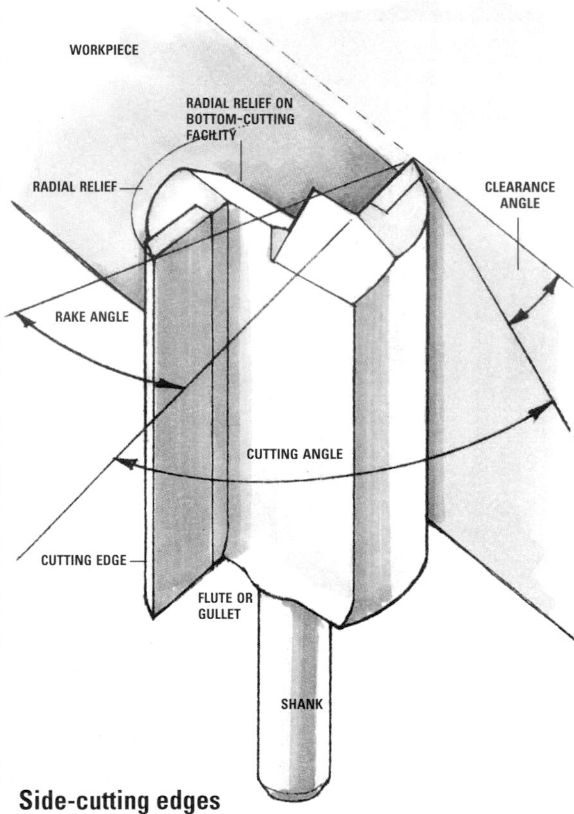

Side-cutting edges

Each side tip, or cutting edge, must be ground to exactly the same angle, size and shape. If one cutting edge is smaller than the other, it will not touch the face of the work, reducing cutting efficiency (see page 27).

Bottom-cutting edges

For plunge-cutting, it is essential that the bottom of the cutter is ground to the same criteria as the sides. This is far more difficult, due to the bottom's smaller area; the combination of deep flutes on the sides and bottom can lead to very weak tips.

The best cutters for specialized boring and plunge-cutting applications are made with a separate tungsten-carbide tip brazed to the bottom of the cutter (see page 18). Cutters without a bottom-cutting facility must only be fed into the work from the side or edge.

Rake angles

Cutters are manufactured with different rake angles – the angle at which the face of the tip meets the surface of the wood. A steep rake angle gives a fast cutting action and removes material quickly. A lower or negative angle produces a slower scraping action, and is mostly used to trim laminates and finish edges.

Clearance angle and radial relief

The clearance angle is the angle immediately behind the cutter tip; radial relief is the angle over the remaining back face of the cutter. Both are ground to prevent the back of the cutter (the surface behind the cutting edge), or waste deposits adhering to it, from rubbing the face of the wood.

Shank diameters

Cutter ranges are available with imperial shank diameters of ¼in, ⅜in and ½in, designed to fit appropriate-size collets (see page 11). Cutters are also manufactured with shanks in metric sizes, typically 6mm, 8mm and 12mm. Imperial and metric shank sizes are not interchangeable – it is not possible to fit an imperial shank in a metric collet, and vice versa.

CUTTER BALANCE

The cutter must be dynamically balanced, to ensure that it rotates concentrically to the spindle axis. If it does not, it will set up vibrations in the router, producing a poor finish on the work and possibly resulting in damage to the cutter and router.

RIPPLES CAUSED BY UNBALANCED CUTTER

Plain or fence-guided cutters

These cutters are generally guided through the work by running the router base, or a fence fitted to it, against the edge of the work or against a straightedge clamped to it. To cut shaped or curved work, you can either use guide bushes to follow the edge of a pre-cut pattern or template, or use a trammel rod to scribe circles and regular curves (see page 54).

Self-guiding cutters

These cutters are guided by a solid pin ground on the end of the cutter, or by a bearing race fitted either on the end of the cutter or on the cutter shank, above the cutting edges. You can then run the pilot pin or bearing against the straight or curved edges of the work, or against the edge of a pattern or template (see page 56).

BUYING CUTTERS

It is difficult to recognize the true quality of a cutter just by looking at it, whatever the price. For this reason, buy from suppliers that offer both a warranty and a reliable back-up and sharpening service.

- *Check that the cutting edges are sharp and unchipped and that there are no voids in the brazing.*

- *Applied tungsten-carbide tips should not overhang the flute other than by the ground bevel, as this may cause them to break off while cutting.*

- *The surface of the cutter should not be pitted, and the cutting edges should be smooth and polished.*

- *Check that the shank is unmarked and is straight and parallel to the cutter axis.*

- *The cutter should run smoothly, without creating any vibrations. If it does not, replace it immediately.*

- *On bearing-guided cutters, the mounting pin or spigot must be strong enough not to snap off under load. You must be able to tighten the holding screw sufficiently, without stripping the thread, to prevent it undoing when routing.*

CUTTER MATERIALS

Most router cutters are made from high-speed steel (HSS) or are tungsten-carbide-tipped (TCT). Because their manufacturing process is simpler, HSS cutters are less expensive than the TCT equivalents. However, a good-quality HSS cutter may be just as efficient and durable as an inexpensive TCT one.

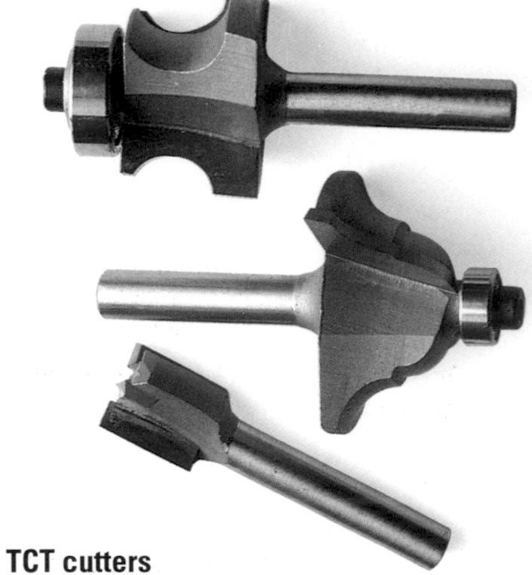

TCT cutters
TCT cutters, which have separate tips brazed onto a pre-ground steel blank, are adequate for most routing applications if kept in good condition. They retain their edge far longer than equivalent HSS cutters, but are more difficult to sharpen and are best honed on a fine diamond or ceramic stone (see page 27). Use only TCT or STC cutters (see below) on resin-bonded and other abrasive materials, and on difficult or abrasive woods.

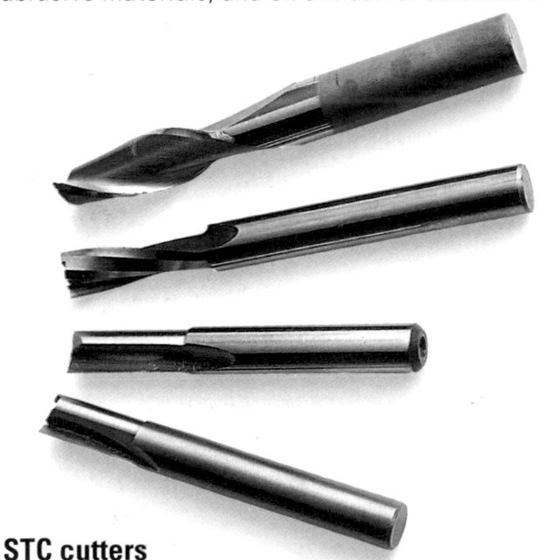

STC cutters
Solid-tungsten-carbide cutters (STC) are generally small in diameter, because the material is costly, and it is difficult to braze carbide tips onto a small steel blank.

HSS cutters
HSS cutters are machined from solid high-speed steel, with cutting and clearance angles ground directly into the steel blank. You can hone them to a very keen edge on an ordinary oilstone, and they leave a superb finish on good-quality softwood and non-abrasive hardwood. HSS cutters are ideal for fine, detailed mouldings and precision dovetail-jointing, as well as other similar applications in quality cabinet work.

HSSE cutters
Special high-speed steel (HSSE) cutters are produced for cutting, drilling and slotting various grades of non-ferrous metal, uPVC plastic and dense hardwoods. The cutters produce a clean cut and are resistant to abrasive anodized finishes. A coolant or lubricant is used when cutting, and the work must be held securely in position.

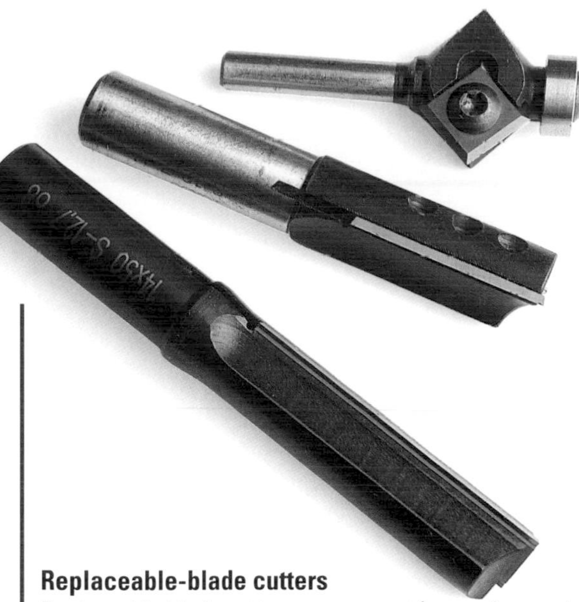

Replaceable-blade cutters

Replaceable-blade cutters are a recent innovation, and consist of a central cutter fitted with removable STC blades. The blades, held by small screws, are honed on all four edges, so you can turn them quickly through 180 degrees once the first pair of cutting edges have been dulled by use.

Angled replaceable-blade trimming cutters are particularly useful for trimming resin-bonded materials, which quickly dull any cutting edge (see page 23).

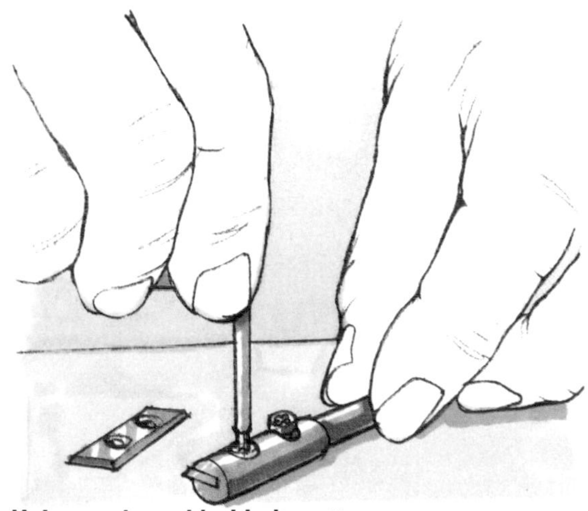

Using replaceable-blade cutters

Replaceable-blade cutters are designed for professional use in fixed overhead or inverted routers. The blades can be rotated or replaced on the bench or, to avoid disturbing the set-up during a production run, without removing the cutter from the router. For plunge-cutting, replaceable single-flute cutters work better than two-flute versions.

CUTTING SPEEDS

Two factors govern cutting speed: the number of cutting edges and the speed of the router motor.

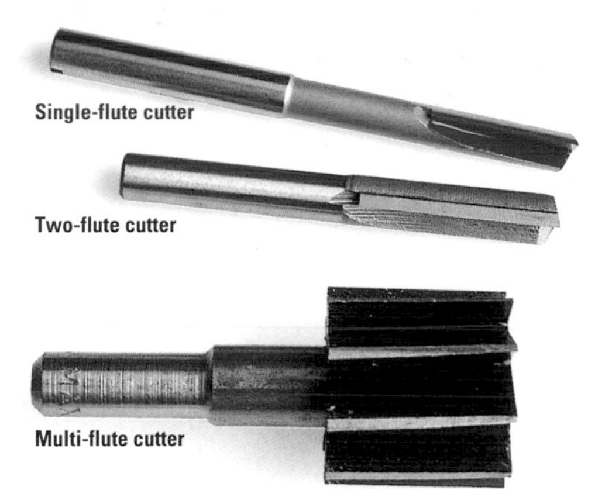

Single-flute cutters

Single-flute cutters, which can be fed into the wood at a fast rate, are generally used for roughing out, because the slower cutting speed (as opposed to feed rate) leaves a less-smooth surface finish.

Two-flute cutters

Two-flute cutters, which are used for most woodworking operations, have a cutting edge ground on each tip, doubling the cutting speed; a cutter used at a motor speed of 22,000 rpm produces 44,000 cuts per minute.

Multi-flute cutters

Multi-flute cutters work in a similar fashion to burrs or rotary files, and are used with low-speed motors (around 3000 rpm maximum). They have poor waste clearance, and must be fed relatively slowly into the work. Do not use these cutters in high-speed routers.

Safe working speeds

Most cutters work safely at maximum router speeds (20,000 to 30,000 rpm). However, cutters over 40mm (1⅝in) in diameter should only be used in inverted or overhead router tables, as the peripheral speed is excessive and can cause a hand-held router to become unstable and unsafe.

As a general guide to speed, do not use cutters with a diameter of 50 to 63mm (2 to 2½in) at more than 18,000 rpm; those of 63 to 75mm (2½ to 3in) diameter at more than 16,000 rpm; and those of 75 to 89mm (3 to 3½in) diameter at over 12,000 rpm. Follow the manufacturer's recommended operating speeds at all times.

STRAIGHT CUTTERS

Straight, or parallel-flute, cutters are most often used for cutting grooves, slots, recesses and housings. You can also employ them for mortising, edge-trimming and template work.

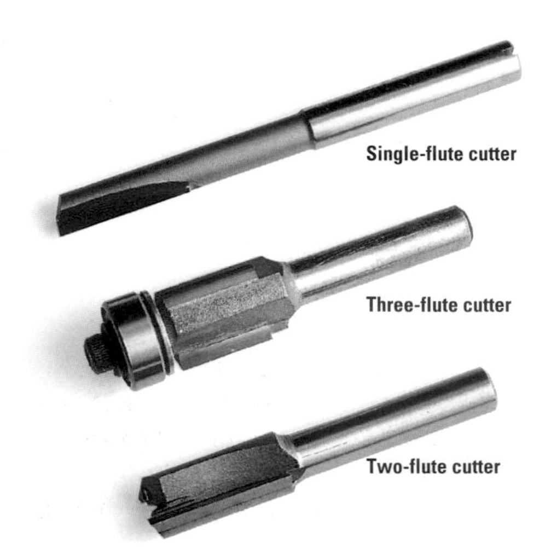

Single-flute cutters
Single-flute straight cutters are best used for fast cutting. The single flute clears waste easily, and is good for plunge-cutting. Small-diameter cutters are ideal for cutting out prior to edge-trimming or profiling.

Two-flute cutters
Two-flute straight cutters are used for most routing operations, including edge-trimming, rabbeting and cutting housings. They produce a superior surface finish to single-flute cutters; their waste clearance is less efficient, however, so use a slower feed rate to avoid overheating. Most two-flute cutters have a bottom-cutting facility for plunge-cutting (see page 16).

Three-flute cutters
Three-flute straight profiles are used as fine-trimming cutters that remove very fine shavings, or as large-diameter cutters with gullets unlikely to hold waste.

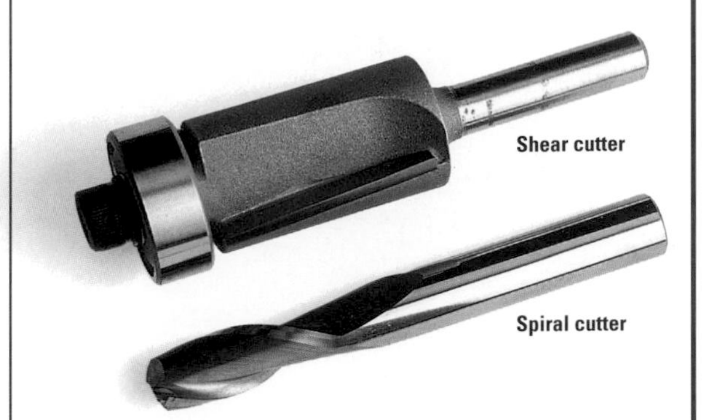

Shear cutters
Shear cutters have slightly angled cutting edges that produce a downward slicing or paring action. This reduces break out along the edges of the cut and leaves a neat, sharp edge. Shear cutters are ideal for veneer-trimming, as they make a neater edge than produced by the 90-degree impact cut of a vertical-edge cutter.

Spiral cutters
These cutters have one or two spiral flutes, similar to a drill bit, and are ideal for small mortises and for cutting pockets. They are either up-cutting or down-cutting (although a few cutters can do both operations in one pass): up-cutting spirals lift the waste out of the cut and, as with shear cutters, the slicing action of down-cutting spirals leaves a neat, clean edge.

Spiral cutters, manufactured for cutting aluminium and other soft metals, are similar to woodcutting spirals, but have more sharply defined flutes.

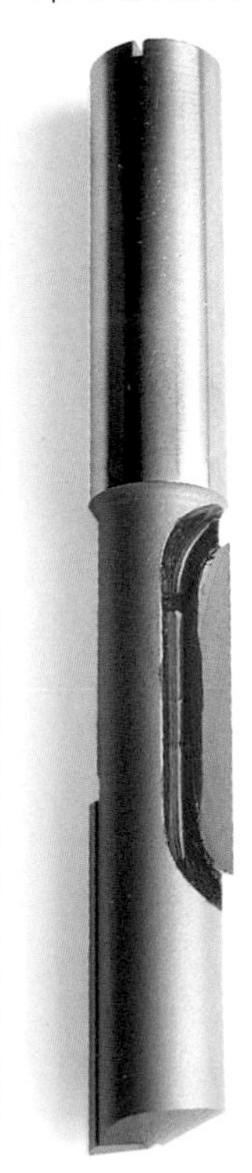

Mortise cutters
Large-diameter mortise cutters are utilized for heavy plunge-cutting when forming mortise-and-tenon joints, deep pockets for locks, and door and joinery fittings. The strong, staggered tips provide good chip-clearance and can be used for bottom plunge-cutting in the same way as single-flute cutters. Although mortise cutters are fairly robust, make cuts in several stepped passes to prevent jamming the cutter in the work.

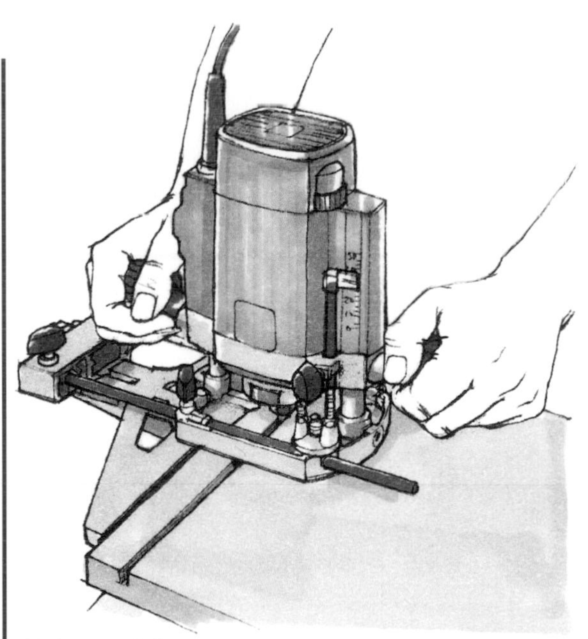

Using straight cutters

Always cut in stages, to avoid overheating the cutting edges or scorching the surface of the work. When using light- or medium-duty routers and cutters with 6 or 8mm (1/4 or 5/16in) shanks, plunge straight cutters in steps of no more than 3mm (1/8in). Increase the depth by the same amount on each pass, up to the full depth.

Cutting plastics

Working carefully, you can cut rigid plastic and acrylic materials with most straight two-flute, spiral and shear cutters. (Specially ground cutters are also available.)

Before routing, spray a non-stick dry lubricant (PTFE) along the cut line, to prevent debris sticking behind the cut and to the cutter itself. Select a low cutting speed, to keep the edge temperature down and prevent shattering along the edges of the cut.

PURPOSE-MADE SADDLES

Using purpose-made fences and jigs to support and guide the router, you can use straight cutters to shape chamfers and bevels or cut mortises and grooves at an angle to the surface of the wood or in a round workpiece. When using a jig where a cutter passes through a guide, make sure the slot or hole provides adequate clearance for waste to escape quickly.

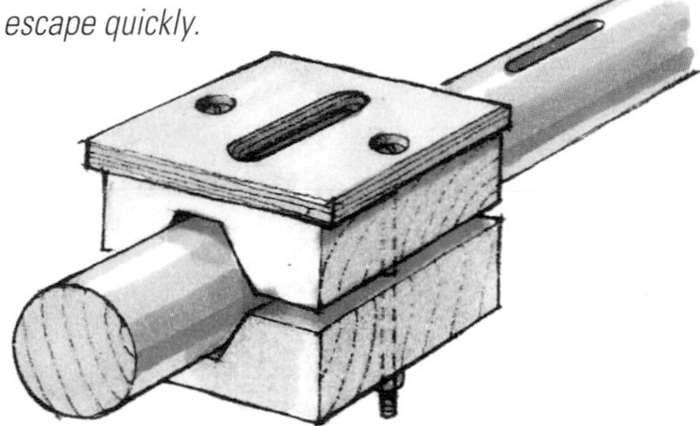

Cutting a mortise in a round leg

Make a saddle, using softwood blocks and MDF, and clamp it over a round leg. Plunge the straight cutter through the slot cut in the MDF board; the slot also serves as end stops to limit the length of the mortise.

Routing slots across a square stile

Make a similar saddle to cut slots across the corners of a square stile to take glass display shelving.

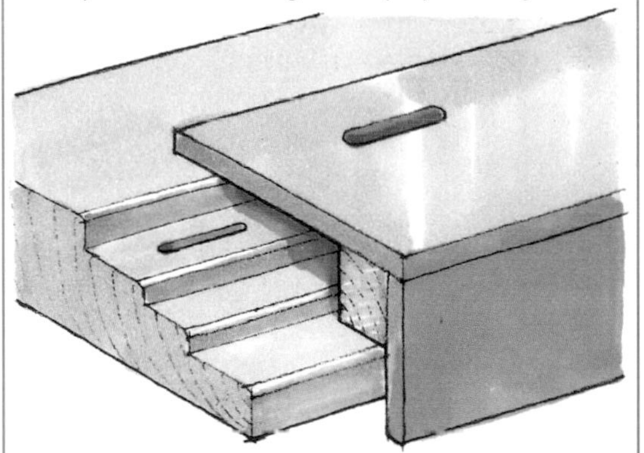

Spanning an irregular surface

An even simpler saddle can be used to support a router over an irregular surface, so that you can machine a stopped groove with a straight cutter.

TRIMMING CUTTERS

Trimming veneers or face laminates with a high-speed router not only ensures that the edges are finished square to the face of the workpiece, but virtually eliminates the risk of chipping or break out. Two- and three-flute trimming cutters are available; the three-flute version produces a more precisely finished edge.

Most trimmers are bearing-guided cutters. When using trimming cutters without guides, utilize the router's side fence.

Use the cutter to skim a fine shaving from the edge of the laminate and across the full thickness of the base material, or to trim the laminate flush with the pre-finished edge of the base material.

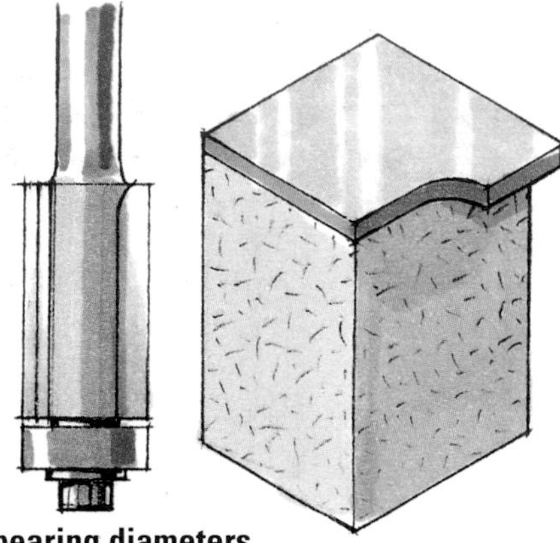

Guide-bearing diameters
Use a guide bearing with the same diameter as the cutter for flush-trimming. For precision work, use one a fraction larger than the cutter, to avoid scoring the finished edge of a workpiece.

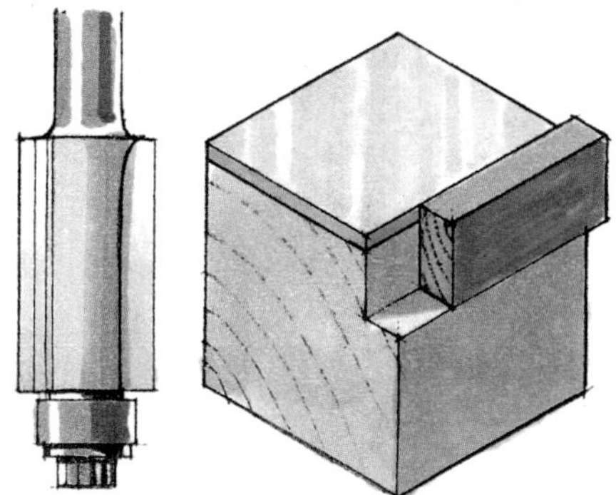

Using small-diameter bearings
Using a bearing smaller than the diameter of the cutter, you can simultaneously trim and rabbet the edge of a workpiece to accommodate a decorative or protective edge bead.

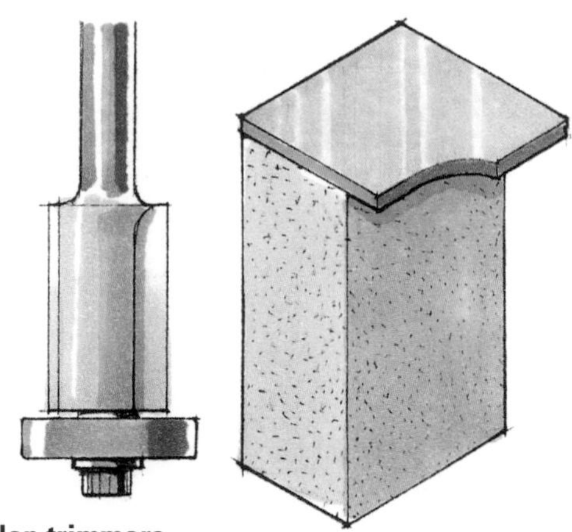

Overlap trimmers
Plastic laminates and other abrasive facing materials dull cutting edges quickly. To extend the life of your trimming cutters, use a special overlap trimmer, a two-flute cutter fitted with an oversize bearing, to pre-trim the overhang down to 3mm (⅛in).

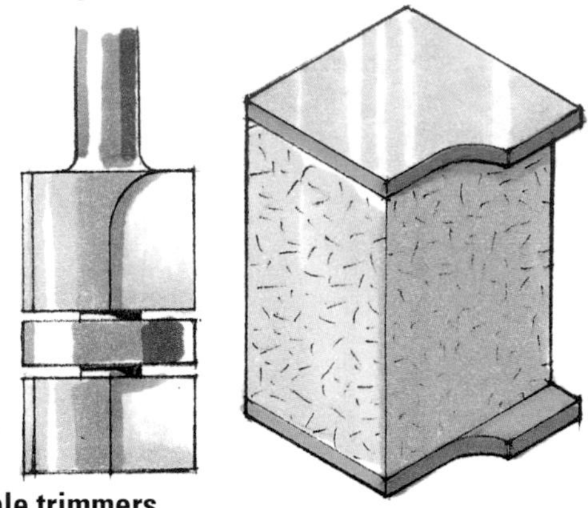

Double trimmers
You can trim the edges of face and balancing laminates in one pass, using a double trimmer. It is made with a roller bearing sandwiched between a pair of cutters, one ground on the shank and the other mounted below the bearing.

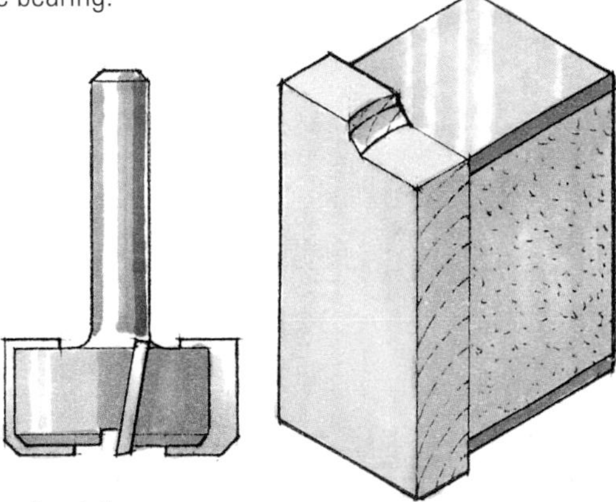

Lipping trimmers
Trim decorative edge lippings flush with a three-flute cutter. Some lipping trimmers have cutting edges with bevelled corners to prevent scoring the work surface.

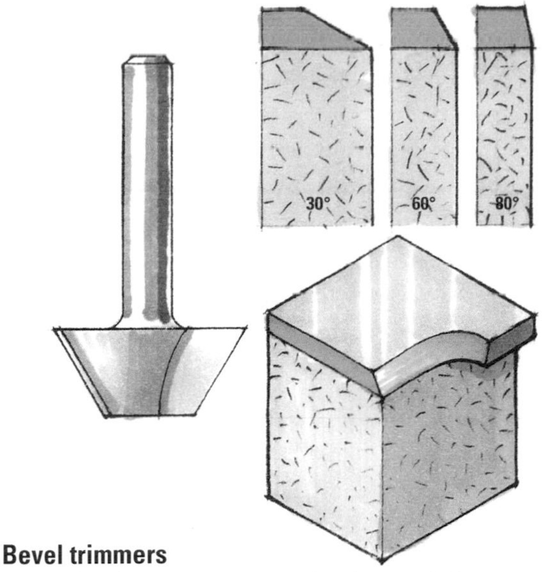

Bevel trimmers

Bevel-trimming cutters remove the sharp arris from laminate edges. The cutters, ground at angles between 30 and 80 degrees, produce a wide, decorative edge line. You can buy plain or self-guiding bevel trimmers.

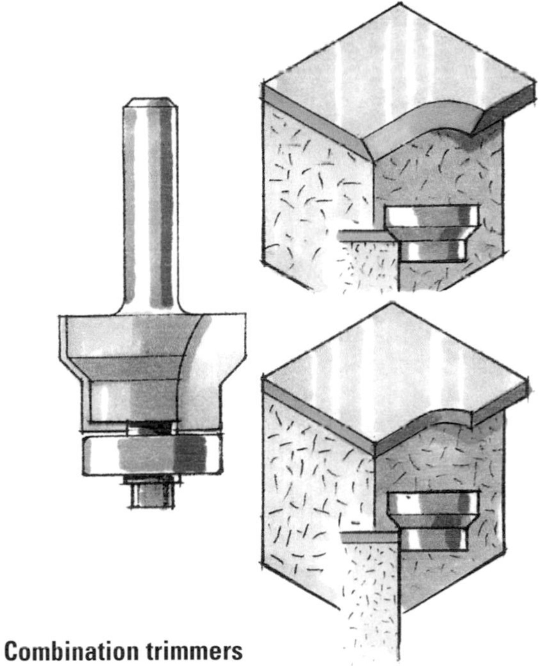

Combination trimmers

Combination trimmers, which are sometimes fitted with guide bearings, have cutting edges ground at a specific angle and parallel edges for flush-trimming. Raise or lower the cutter until it produces the required edge profile. Don't use combination trimmers for chamfering the edges of solid timber, as they are not ground with the appropriate clearances; chamfer or V-groove cutters are better for this operation (see page 25).

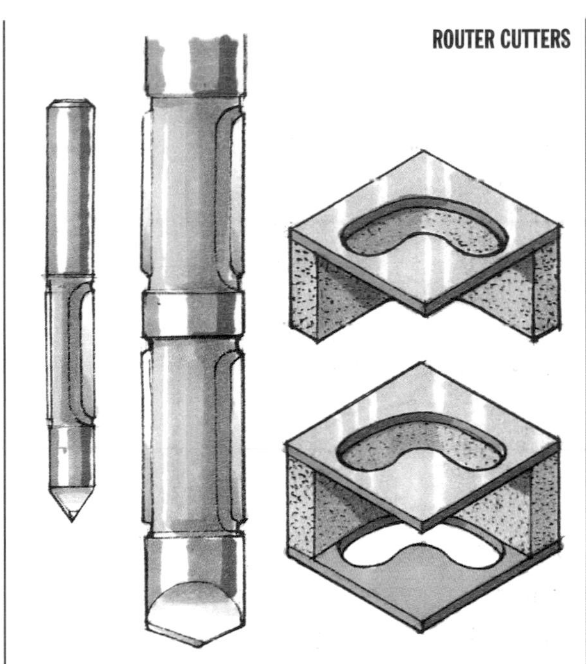

Pierce-and-trim cutters

Pierce-and-trim cutters form internal cutouts for inset fittings, such as sinks and hobs. Cut the opening in the worktop core before applying the face laminate or veneer. Pierce through the unsupported laminate with the drill point of the cutter, then run the pin guide around the internal edge of the opening. Double-trim cutters, for use with heavy-duty routers, cut top and bottom laminates simultaneously.

REMOVING EXCESS ADHESIVE

Let laminate or veneer adhesive dry thoroughly before using bearing-guided trimming cutters, otherwise the bearing will clog and be unable to turn. In addition, remove any glue deposits from the edge of the base material and the overhanging edge of the laminate; if you don't, the bearing, side fence or other guide will be deflected, leaving an uneven edge.

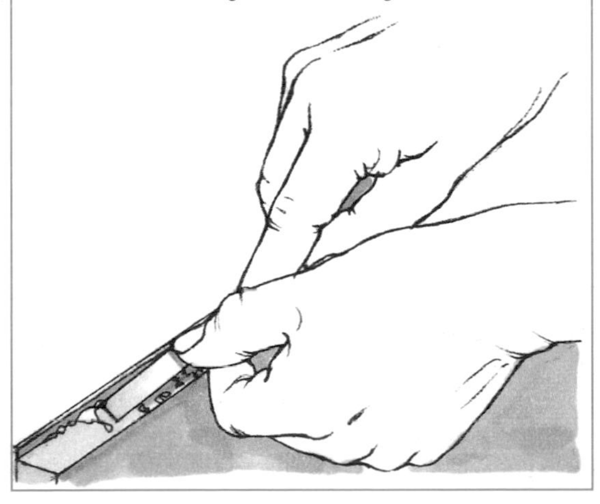

MOULDING CUTTERS

Mouldings have many applications in furniture making and joinery. You can use them as pure decoration, or to visually alter proportion, to make open joints appear more attractive, and to accentuate specific details.

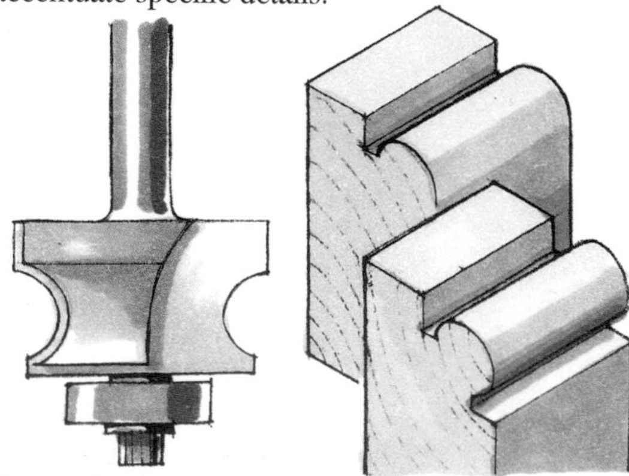

Corner-bead cutters
A corner-bead cutter has the bead profile set low on the cutting edge to leave a narrow bottom edge. Bead mouldings are frequently used to disguise the joint line around frames or linings. For example, a moulded stop bead is employed to surround and hold a traditional sliding-sash window in its frame. Bead mouldings can also be used to provide a decorative break between rabbeted meeting stiles of double doors. To form a complete corner bead, cut along one face of the work before turning the workpiece to make a second pass along the adjacent face.

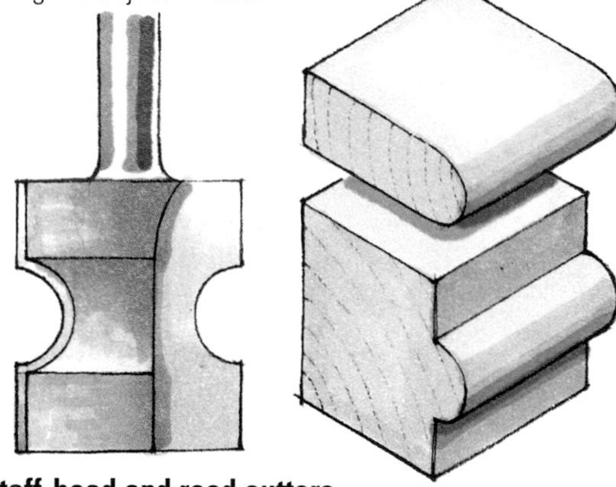

Staff-bead and reed cutters
Staff-bead and reed cutters make decorative edge mouldings, typically for table tops and chair legs. You can also use a staff-bead cutter to make simple parting beads for sash windows, or use one in combination with a matching sunk-bead cutter to make interlocking edge joints (see page 104).

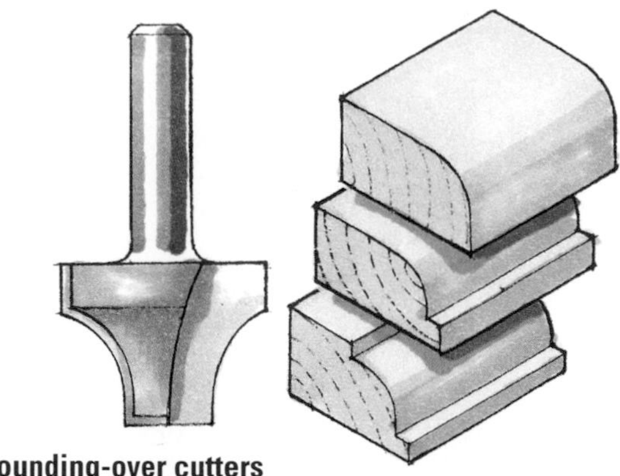

Rounding-over cutters
A rounding-over cutter, with or without a guide bearing, is used to radius the arris of a workpiece. By varying the router settings, it is possible to form a quirk or step on one or both edges.

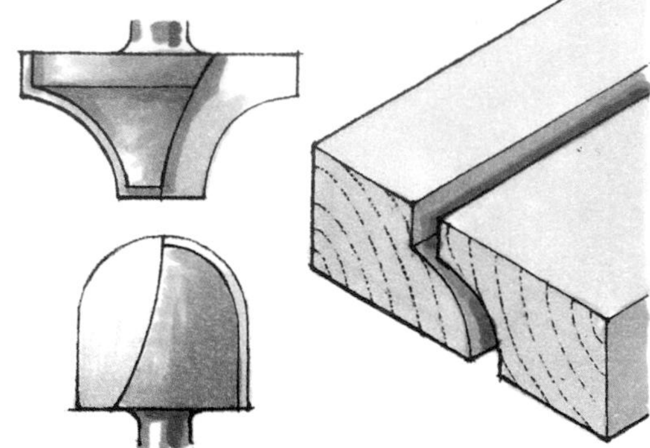

Reverse-profile cutters
Matched pairs of moulding cutters form a moulding and its reverse profile. For example, you can cut traditional rule joints for drop-leaf tables using matching cove and rounding-over cutters. A glazing-bar-moulding cutter and its matching reverse-scribing cutter produce rails, stiles and sash bars for window frames.

PRODUCING COMPOUND MOULDINGS

Use combinations of decorative-moulding cutters to vary a routed profile or create compound mouldings.

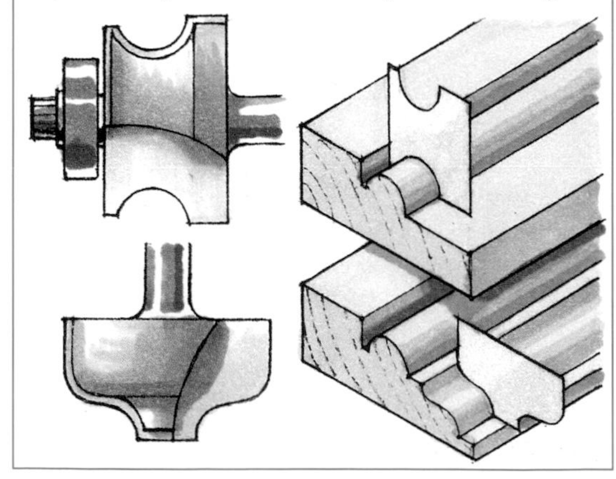

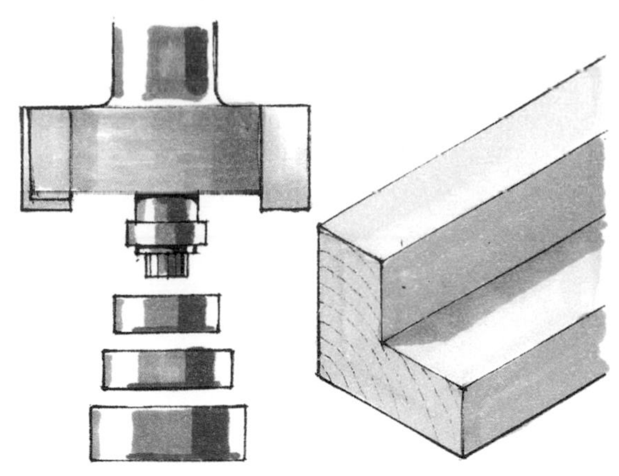

Rabbet cutters

Use a rabbet cutter and a guide bearing to router rabbets and recesses along the edge of a workpiece. Rabbet cutters have cutting edges on both the side and bottom of the tips. The depth of the rabbet is determined by how far a cutter can project from the base of the router. In general, a cutter with a 6mm (¼in) diameter shank will cut rabbets up to around 12mm (½in) deep. Maximum rabbet width is the overall radius of the cutter minus the outside radius of the guide bearing. Fit a different guide bearing to alter the rabbet width.

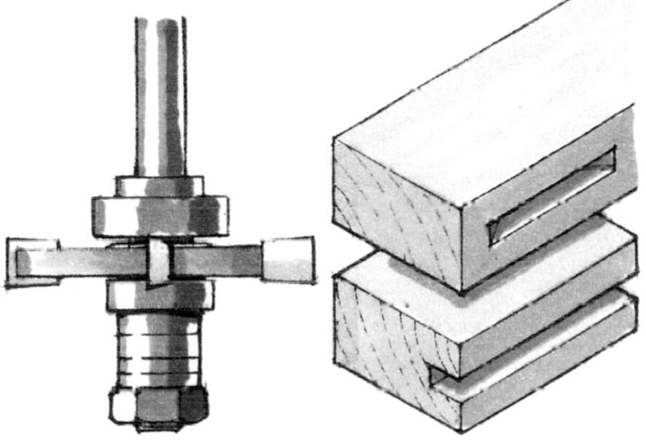

Arbor-mounted slotting-and-grooving cutters

Slotting-and-grooving cutters, mounted on a threaded arbor, consist of thin steel discs fitted with one or more tungsten-carbide teeth. They are available in various 'blade' thicknesses, and are mainly used for cutting slots or grooves along edges.

Cutters are available in diameters of up to 100mm (4in); however, do not use cutters larger than 50mm (2in) in diameter in hand-held routers. Assemble the cutters carefully, checking that you have fitted the blades the correct way round and have firmly tightened them, to prevent the arbor turning in the bore.

You can use shims, spacers, collars and guide bearings to assemble arbor-mounted cutters to very precise tolerances, and to vary the edge profile and guide-edge position.

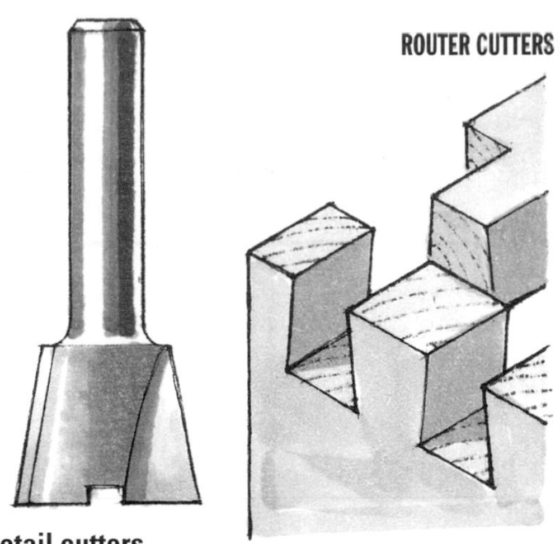

Dovetail cutters

Dovetails are utilized in all forms of furniture construction. Dovetail cutters also form dovetail-profile housings and matching tongues.

Use dovetail jigs and cutters to make sets of through and lapped dovetails for drawers and boxes, as well as more-complex mitred and hidden dovetails.

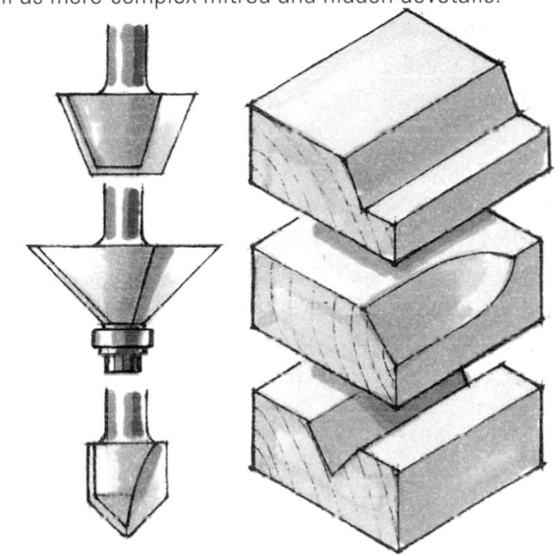

Bevel, chamfer and V-groove cutters

You can use bevelled self-guided or fence-guided cutters to machine decorative chamfers, stopped chamfers and bevels along workpiece edges. These cutters are made with bevel and chamfer angles of between 25 and 60 degrees. For stopped chamfers, clamp stops at each end of the cut, to limit the extent of the cutter travel.

Using deep-bevel cutters of up to 40mm (1⅝in), you can machine tapered-face architraves and skirtings, as well as cutting the angles for segmented-construction boxes, and pilasters and columns for faceted and turned work. Use these larger cutters only in table-mounted routers.

Small V-groove cutters are used for relief carving and decorative or letter engraving.

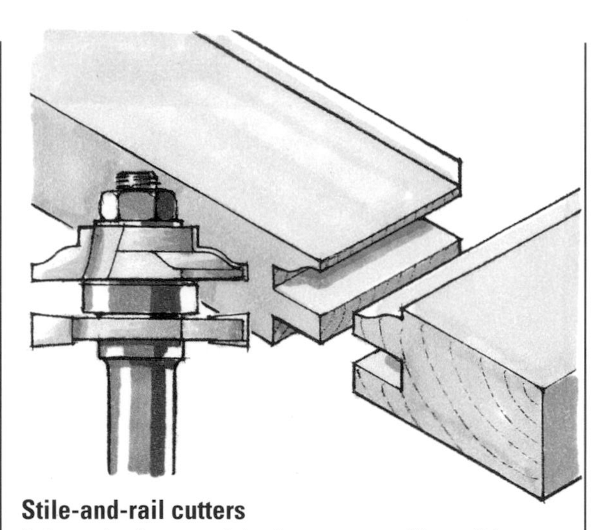

Stile-and-rail cutters

Stile-and-rail cutters (also known as profile-scribing sets) are designed to produce the moulded frames for panelled cabinet doors. A pair of cutters, separated by a bearing, shapes the rail-end scribed moulding and a short tenon. The same cutters, reassembled on their arbor, are then used to shape the matching panel groove and edge moulding. When you glue the tenon into the panel groove, the scribed moulding allows the rail end and stile edge to butt together. You can also use stile-and-rail cutters to make straight or curved top rails for doors .

Stile-and-rail cutters are suitable only for heavy- or medium-duty routers mounted on an inverted or overhead table. When cutting the groove and moulding on curved rails, use workholders to control the work (see page 75).

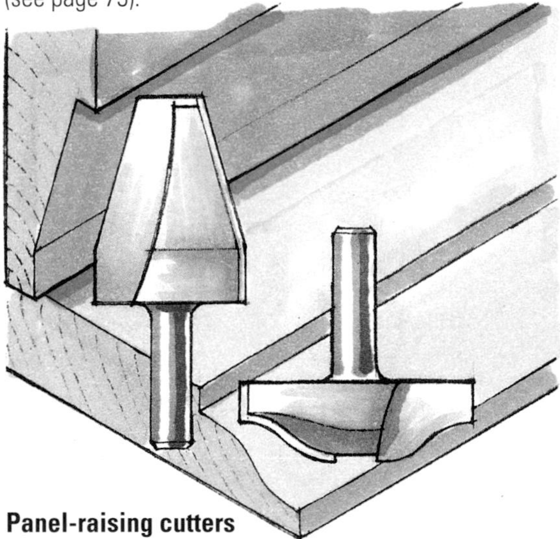

Panel-raising cutters

These cutters, used for making traditional raised-and-fielded panels, are among the largest cutters available. There are two types, large-diameter horizontal cutters and smaller-diameter vertical ones; the horizontal cutters must be used in a fixed router only, but you can employ the vertical cutters in a hand-held router.

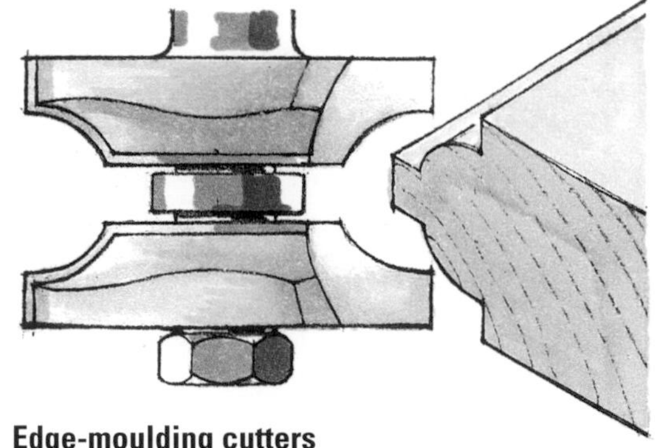

Edge-moulding cutters

Edge-moulding cutters profile top and underside workpiece edges in the same pass. Either run the workpiece against a table fence, or guide a hand-held router, using the side fence or a guide bearing fitted between two cutters mounted on a threaded arbor.

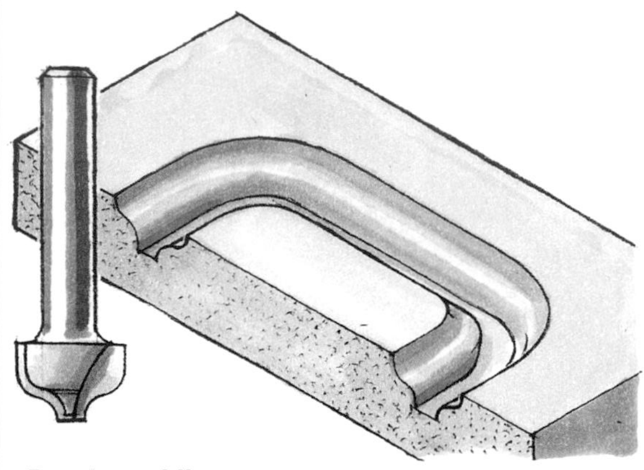

Panel-moulding cutters

These cutters simulate traditional panelling by machining the profile from solid-wood or man-made-board cabinet doors and drawer fronts.

WORKING WITH MOULDING CUTTERS

Although you may be able to make relatively deep cuts on the first few passes, reduce the amount of material being removed as you bring the full length of the cutting edges into contact with the wood.

MAINTAINING CUTTERS

Contrary to what one might expect, overheating rather than natural blade wear is the main cause of router cutters becoming blunt. Once a cutter has started to dull, any further use results in the steel rapidly losing its tempering and not holding a keen edge. The two most common causes of overheating are waste that packs tightly in the cutter flutes, and a too-slow or inconsistent feed rate. Using a blunt cutter can lead to: poor surface finish; tearing of timber grain; scorching or burning the work; and overloading the router motor and bearings.

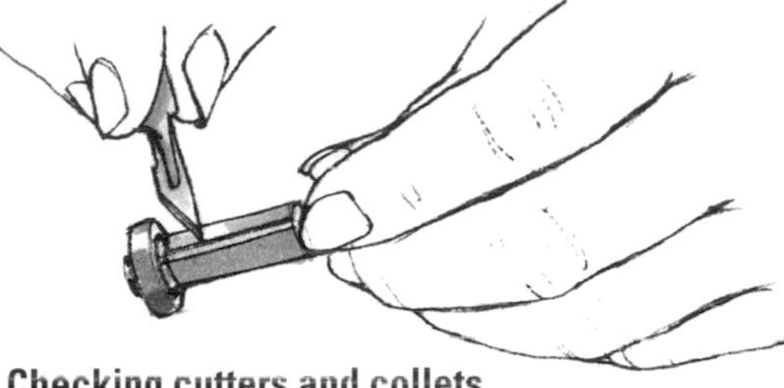

Checking cutters and collets
Regularly check the condition of your cutters and keep them sharp. Remove any traces of resin and sawdust from the cutting faces and from the corners of the flutes. Radial scoring on the shank may mean that the cutter has been slipping in a worn collet.

Worn or damaged collets run cutters out of true, as well as damaging their shanks. Check that the collet surfaces are smooth and free from burrs, and that the nut and thread are in good condition. When tightening a collet, take care not to cross the thread, and do not overtighten the nut.

Storing cutters
Store cutters in a plastic case with foam padding, or in a wooden cabinet with drilled racks. Don't keep your cutters on the workbench or near areas where they can get knocked against other tools. If you do not intend to use router cutters for a while, spray them with a thin oil or rust-protecting agent.

Sharpening stones
To sharpen HSS cutters, use the same oilstones or India stones that are employed to hone chisels and plane irons. TCT cutters must be honed on hard stones, such as diamond-impregnated or ceramic stones, used dry or lubricated with a little water. Keep a pocket slipstone handy for touching up the edges regularly. You can also use these stones to hone HSS cutters.

RESTORING CUTTERS

Badly chipped cutting edges cannot be restored easily, if at all. However, you can sometimes remove small, shallow chips by honing the flat face lightly. In some cases you can have cutters professionally reground, although this is unlikely to prove economical for small or inexpensive cutters.

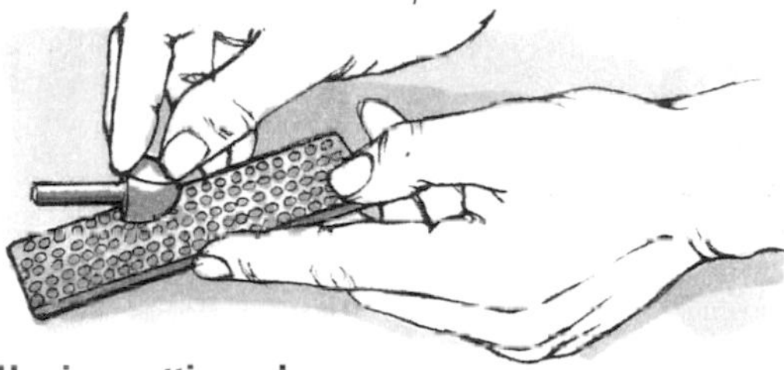

Honing cutting edges
Hone only the flat faces of the cutting edges. Keep each face flat on the stone and rub it back and forth several times. Do not attempt to grind or hone the bevelled cutting edge.When sharpening cutters with more than one edge, hone each one an equal number of times – if cutting edges are not the same profile and size, their effectiveness is reduced.

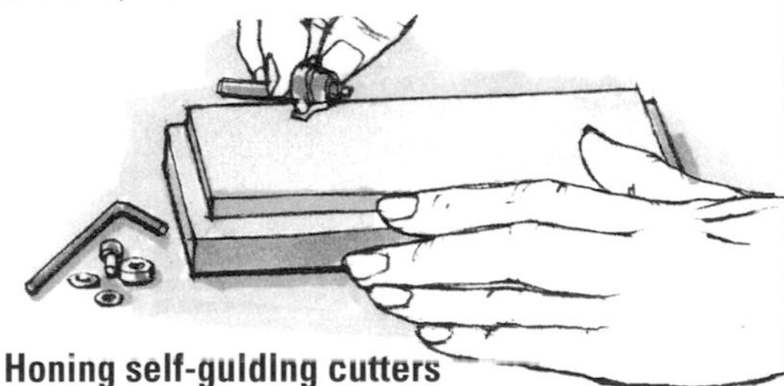

Honing self-guiding cutters
When honing fixed-pin self-guiding cutters, take care not to run the pin against the edge of the stone. Remove the bearings from bearing-guided cutters before honing (see page 52).

ROUTER COLLETS

To hold a cutter in a router, a tapered or cone-shape collet or chuck is fitted to the end of the motor spindle, where it is located in a matching taper machined inside the spindle end. When a locknut is tightened, it compresses slits machined in the collet sides, and these grip the cutter shank. Most collets are manufactured to suit specific shank diameters, but you can also buy collets that take shanks of different diameters. Router collets are generally either of the precision-ground multi-slit or simpler and less-accurate split-cone types.

Multi-slit and split-cone collets
Multi-slit collets grip the cutter with even pressure around the shank, ensuring that the cutter is held firmly in alignment along the centre axis of the spindle. Split-cone collets have fewer slits than multi-slit collets, and are prone to pinching and raising a slight burr on the shank. This can misalign the cutter and cause it to run out of true, setting up vibrations in the router.

Reducing sleeves
An alternative method of holding shanks is to use reducing sleeves. These are an inexpensive way to decrease the collet size; however, they do not always hold the cutter securely, and can damage the shank.

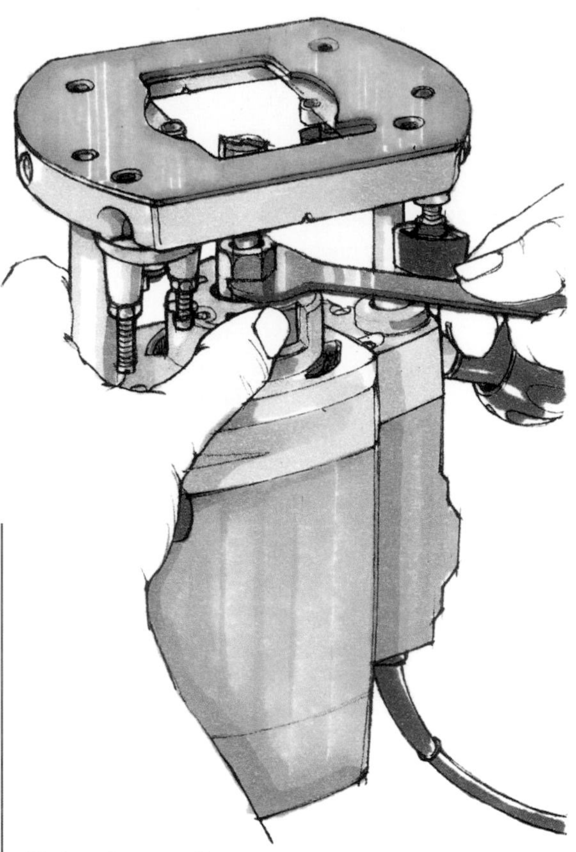

Tightening a collet
Many routers are fitted with a spindle lock to prevent the router spindle turning when you tighten the collet nut. Use only the correct-size spanner when tightening, to avoid damaging the nut. Take care not to cross the fine thread of the collet nut, and do not over-tighten.

A double-lock system is used for safety on some router collets; here, the collet nut tightens, then gives and tightens for a second time. Make sure that the nut is fully tightened before using the router.

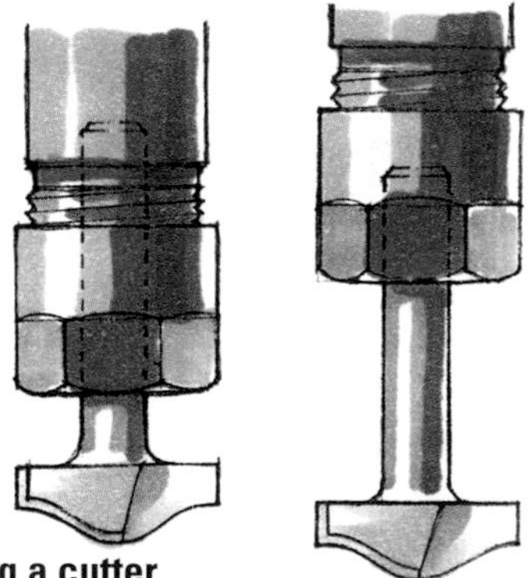

Fitting a cutter
Don't make even fine cuts unless at least three-quarters of the cutter shank is held securely in the collet. If you insert just the tip of the shank, the cutter may come out and spoil the workpiece, and there is also a risk of damaging the collet and the shank itself.

CHAPTER 3 Before getting down to work with a router, it is essential to familiarize yourself with a few basic procedures. These fall into two categories: the first is knowing how to set up and adjust your router; the second is understanding and practising the different ways of feeding a cutter into a piece of work.

INITIAL SETTINGS AND ADJUSTMENTS

In common with all woodworking machines and power tools, an electric router must be carefully set up, checked and adjusted before it can be used successfully. Without following a few simple procedures, you will not be able to produce clean-cut, accurate work. Your first and most important decision should be to buy a good, precision-made router. Inexpensive routers may not always be assembled with care and attention to detail, and may be more likely to develop faults; all efforts to set up such a router accurately will be in vain.

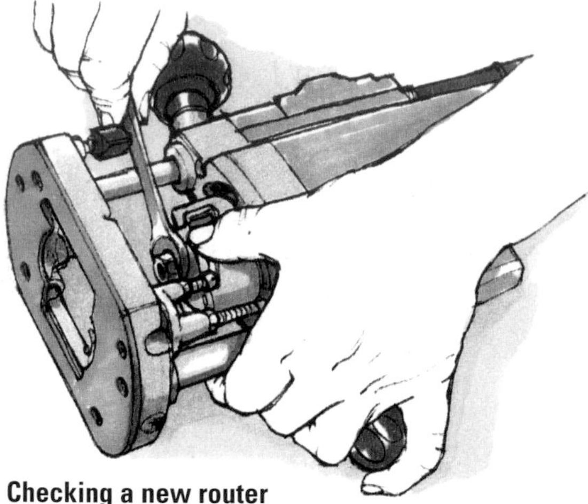

Checking a new router
First remove the collet and locknut, then switch the router on and check that the motor runs without setting up vibrations. Fit a cutter and make a test run on a piece of scrap timber.

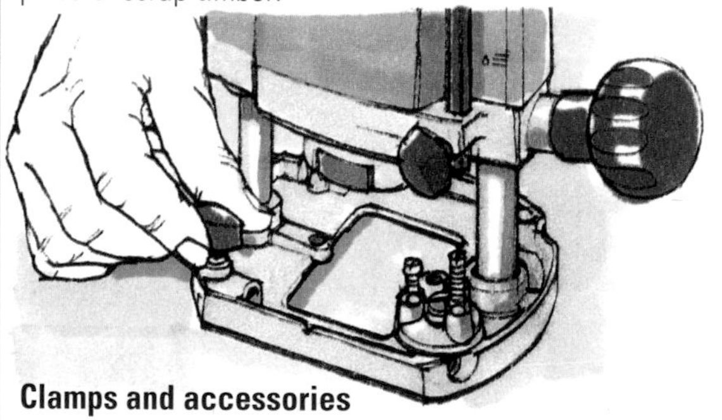

Clamps and accessories
Check that all clamping screws and other threaded parts engage without binding or cross-threading, and that screw threads are not clogged with swarf. Make sure that all separate components and accessories fit together properly, and that the router's fine adjusters and other controls work smoothly.

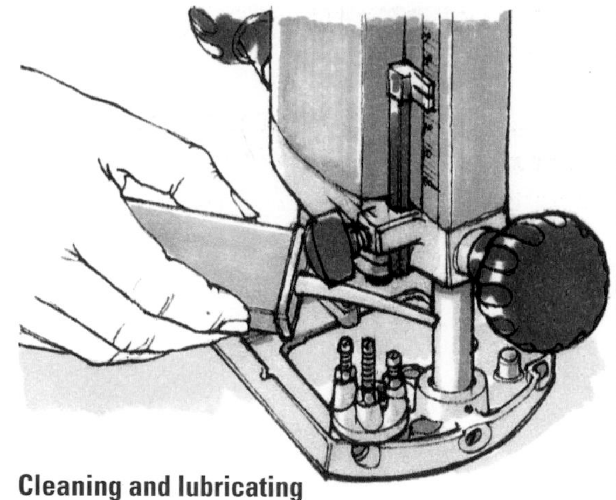

Cleaning and lubricating
Clean any protective grease from the collet, spindle and plunge columns, and then apply a thin smear of light machine oil, to prevent them rusting and to ensure that they operate faultlessly.

CONTROLLING THE ROUTER

When first using an electric router, novices often grip the tool too tightly, pressing it hard down onto the work. However, this tends to inhibit smooth lateral movement, causing the router to snatch. Hold the router by the side handles in a firm but relaxed way, and complete the pass in a steady, continuous movement, maintaining a constant feed rate throughout. Take care not to slow down at corners or any other change of direction, otherwise the cutter can scorch the wood at those points.

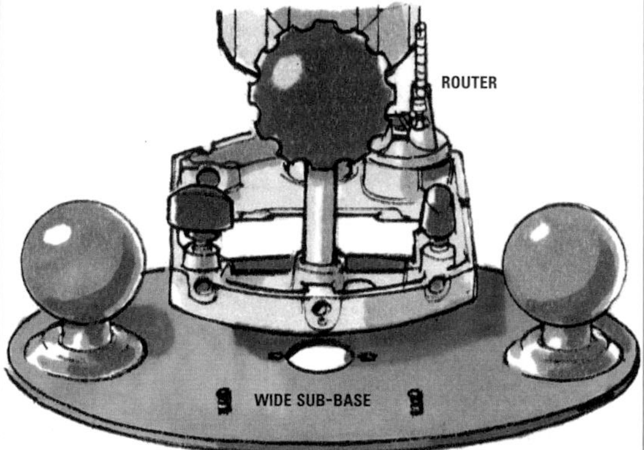

Lateral control
Some plunge, and most fixed-base, routers have side handles set fairly low, for good balance and easy control. For freehand operations, however, it is better to hold the router by the base or, for safety, by side handles mounted on a specially made sub-base fitted to the base. Use one of these methods to reduce any tendency for the router to tip when cutting.

SETTING DEPTH STOPS

All routers are fitted with depth stops, or gauges. These range from a simple rod and clamping screw to a sophisticated geared sliding gauge with a multi-position stop fitted to the base plate. These stops pre-set the depth to which the cutter enters the wood, allowing you to cut grooves, housings and recesses to a consistent depth.

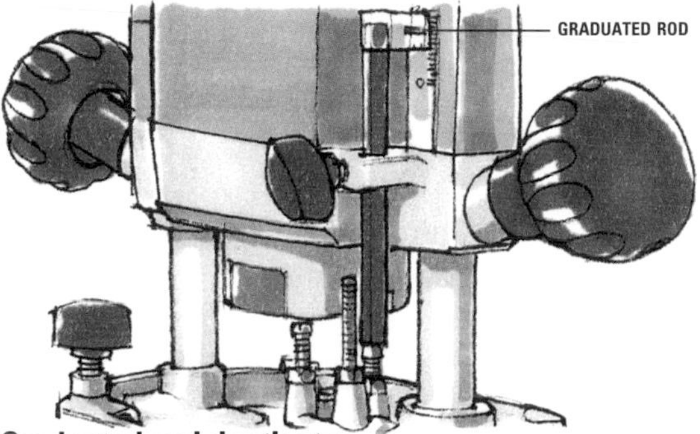

Graduated-rod depth stop
The simplest form of depth stop is a rod set against a graduated scale on the motor housing. The depth of cut is adjusted by reading off and setting the distance between the tip of the rod and its stop on the router base plate, equal to the required depth of cut.

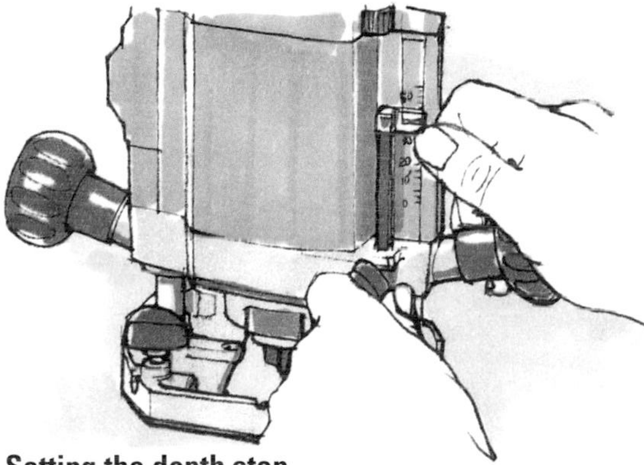

Setting the depth stop
Before setting the depth stop, check that the router is disconnected from the power supply. Fit a cutter in the collet and tighten the locknut.

Place the router on a flat surface and press down on the motor housing until the cutter rests on the surface, then secure it in that position with the plunge lock. Lower the depth-stop rod or slide until it rests against its stop on the router base. Referring to the scale engraved on the rod or slide, raise the rod by the required depth of cut and tighten the depth-rod locking screw. Release the plunge lock and allow the router to rise to its full extent.

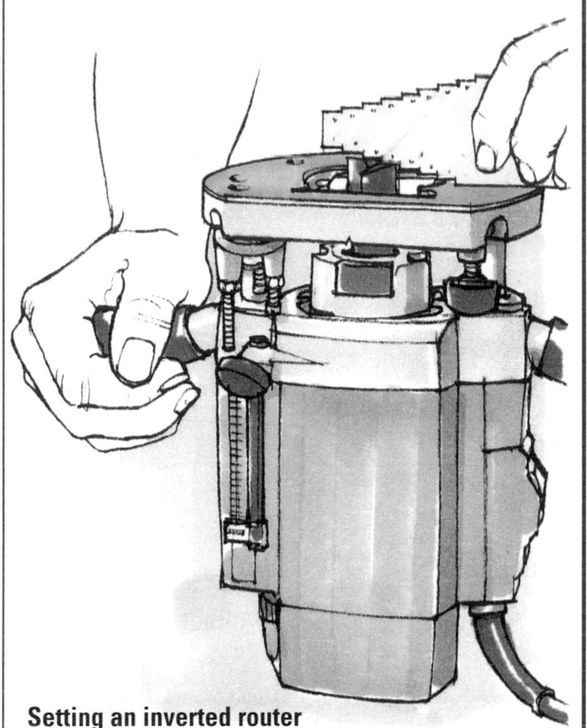

Setting an inverted router
Alternatively, invert the router and operate the plunge mechanism, using a steel rule or gauge to measure how far the cutter protrudes through the base plate. Lock the plunge mechanism, then slide the stop rod up against its stop on the base of the router.

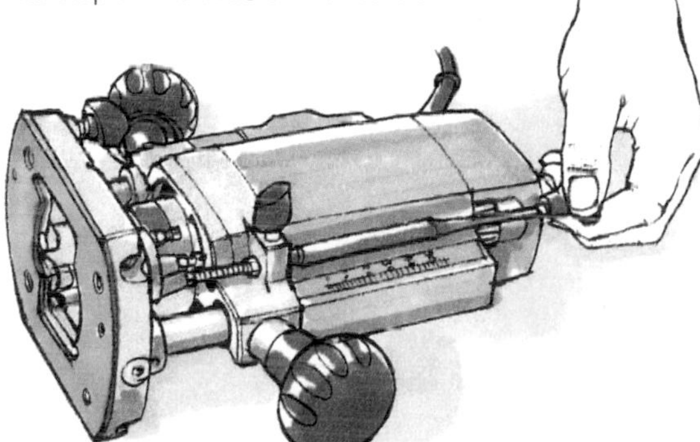

Setting a fine depth adjuster
For very precise depth control, some routers are fitted with a fine adjuster, which either fits over the plunge-return stop rod or replaces the standard depth-stop rod. This adjuster overrides the plunge action and allows accurate screw-feed depth adjustment. Fine adjusters are particularly useful for table-mounted routers where plunge action is not required.

To set the adjuster, lay the router on its side (if it is not fitted to a table or stand), and release the plunge lock. Screw the fine adjuster in or out until the cutter protrudes the required amount from the base, then secure the plunge lock to prevent accidental movement of the setting while working.

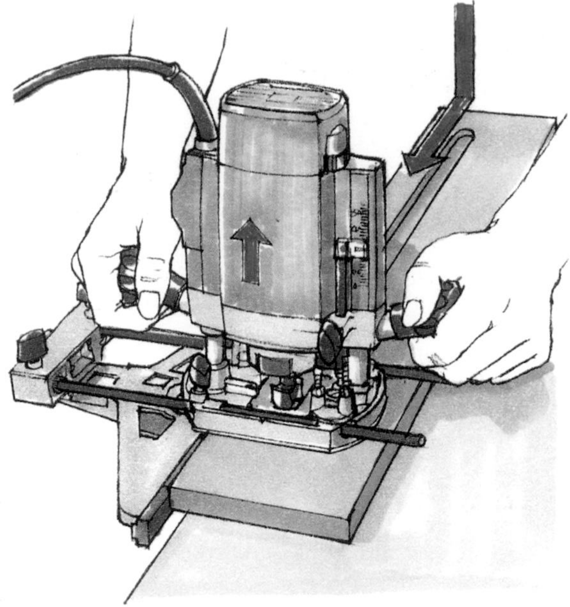

Cutting to a set depth

When starting a cut from the edge of a workpiece, place the lip of the router base flat on the wood so that the rest of the base overhangs. Plunge the router to the set depth and secure the plunge lock before switching on and feeding the cutter into the work.

On completing the cut, release the plunge lock to withdraw the cutter before switching off and lifting the router from the work.

When starting a stopped cut, switch on and plunge down into the wood to the pre-set depth, then apply the plunge lock before feeding the cutter forwards.

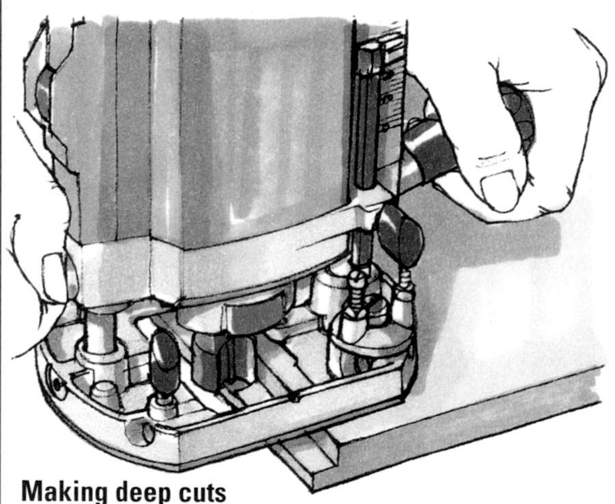

Making deep cuts

To make a deep cut in stages, adjust the depth stop between passes (see above right). Always switch off before making adjustments, and check that the fence or guide is held firmly against the same guide edge when repeating the cutting operation. Continue in stages to the full depth; make a very fine cut of no more than 1.5mm (1/16in) on the final pass, to leave a precise, smooth finish.

OPTIMUM CUTTING DEPTH

During routing, friction generates heat along the thin cutting edges of the cutter. Excessive heat will ruin the temper of the steel and blunt the cutter, and is also likely to scorch the surface of the timber. One way to prevent this is to make relatively deep cuts in stages, rather than cutting to the full depth in one go.

Always check that the depth of cut is set correctly by making a trial cut on a piece of scrap wood.

With light-duty or medium-duty routers fitted with 6mm (¼in) diameter-shank cutters, don't cut deeper than 3mm (⅛in) in one operation. This can be increased slightly, cutting from 4 to 6mm (5/32 to ¼in) deep, when using a router rated at 900 watts and fitted with 8mm (⅜in) shank cutters. Heavy-duty routers fitted with 12mm (½in) shank cutters, will cut from 6 to 8mm (¼ to ⅜in) without overheating.

These guidelines apply to straight, parallel-sided TCT cutters; reduce the cutting depths a little when using HSS cutters. In addition, you may need to vary the cutting depth when working hard, difficult or abrasive materials.

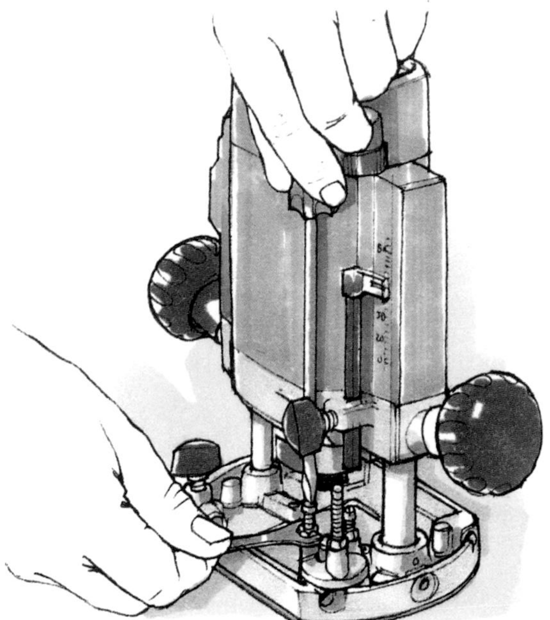

Setting a turret stop

To avoid having to reset between cuts, many routers are now fitted with a turret stop that enables you to pre-set three or more different cutting depths.

To set each adjustable stop on the turret, slacken off its locknut and use a screwdriver to raise or lower the screw. Measure the height of each screw stop, using a steel rule, then re-tighten its locknut.

FEEDING THE ROUTER

As a general rule, the feed direction – that is, the direction in which the router is fed across the work – must be against the rotation of the cutter. This way, not only does the cutter work effectively, but the rotational force also pulls the side fence or guide (self-guiding cutter bearing, guide bush, etc.) against the edge of the work. Feeding in the opposite direction makes the fence or guide wander away from the edge or line of cut.

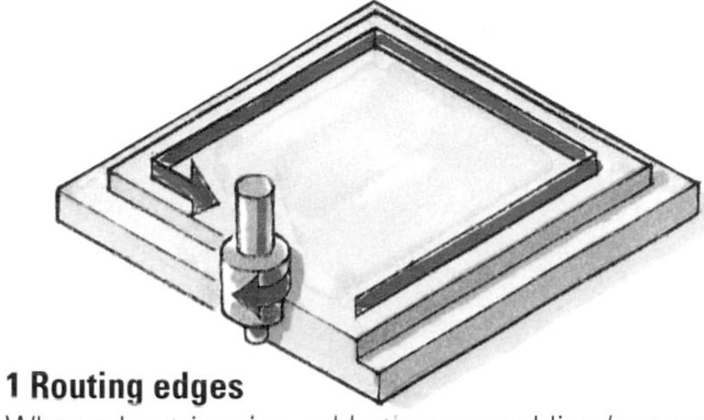

1 Routing edges
When edge-trimming, rabbeting or moulding (see pages 60–1), the feed direction should generally be against the rotation of the cutter.

2 Using a side fence
The feed direction should be against the rotation of the cutter when using a side fence (see page 44), in order to keep the cutter running parallel to the edge, particularly when routing the face.

3 Using a straightedge or T-square
When working against a straightedge, ensure that the feed direction is against the rotation of the cutter, in order to pull the edge of the router against the guide (see page 47).

FREEHAND ROUTING

For freehand routing operations, such as removing the ground when relief carving, feed the router so that it is braced against the pull of the cutter, allowing for any variation in the resistance caused by soft and hard areas of the wood. Maintain a single feed direction as much as possible, however; working from left to right will effectively keep the cutter pulling away from you, which is generally the safest and easiest way to keep control.

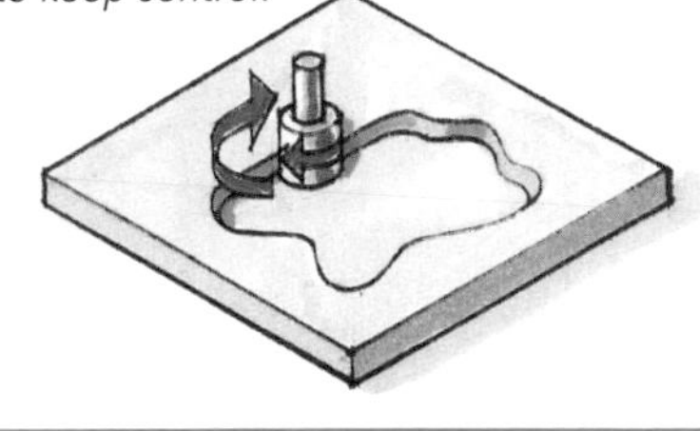

4 Using internal templates
Feed the router clockwise around the inside of a template (see page 50).

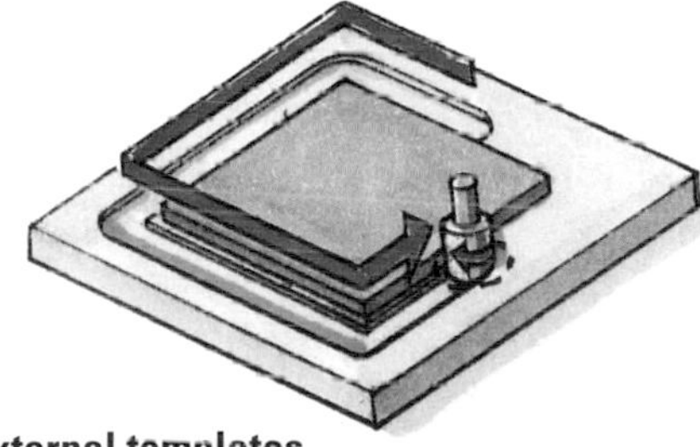

5 Using external templates
When routing around the external edges of a template, feed the tool anticlockwise (see page 50).

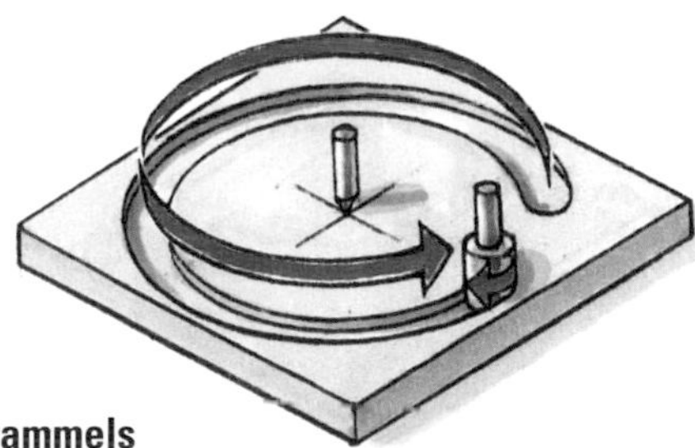

6 Using trammels
When using trammels to router a circle or ellipse, feed the router in an anticlockwise direction, to brace the cutter against the centre point along the trammel rod, rather than away from it (see pages 54–5).

BACK CUTTING

There are exceptions to every rule. When a workpiece is held upright in a vice, feeding the router in the usual direction – against the cutter rotation – tends to damage the wood fibres along the lower edge of the cut. Although a sharp cutter produces minimal break out, the effect is exaggerated when cutting across end grain. Similarly, using a router on veneered or laminate-faced materials can leave them with an unsightly ragged or chipped edge.

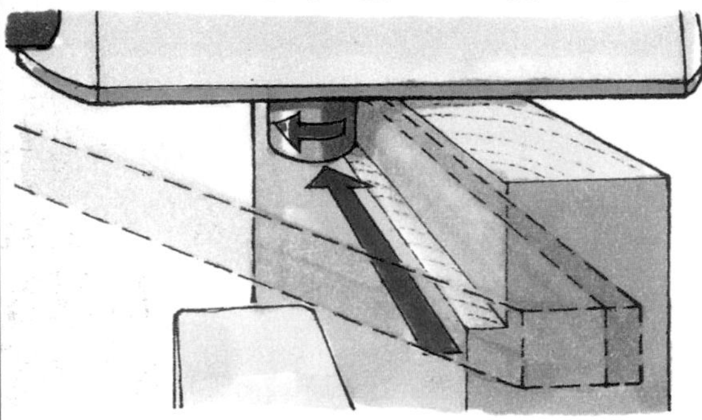

Back cutting end grain
For a really clean-cut edge along end grain, make a very shallow initial cut, feeding the router with the direction of cutter rotation. This will trim the end-grain fibres neatly before you remove the remaining wood by feeding the router against the cutter rotation.

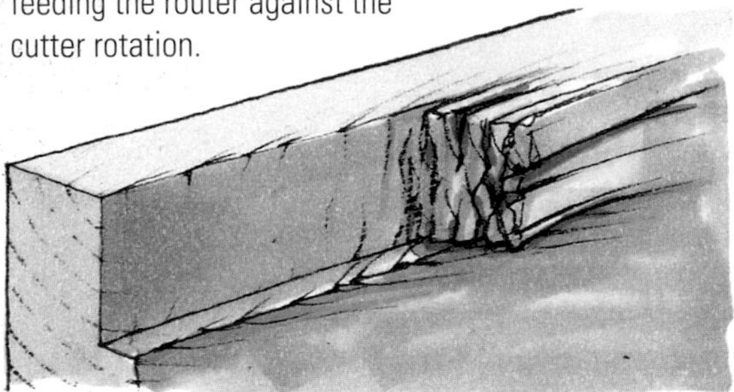

Routing difficult wood
If you are working with a timber that is prone to splitting out as long slivers ahead of the cutter, shallow back cutting may be a solution. But try reducing the depth of cut to very fine shallow steps first, feeding the router against the cutter rotation.

Keeping to the guide edge
When back cutting, it is most important to maintain control over the router throughout, to prevent the fence or guide from wandering away from the guide edge. When using templates, always ensure that if the router should wander off its line, the cutter will be directed into waste material.

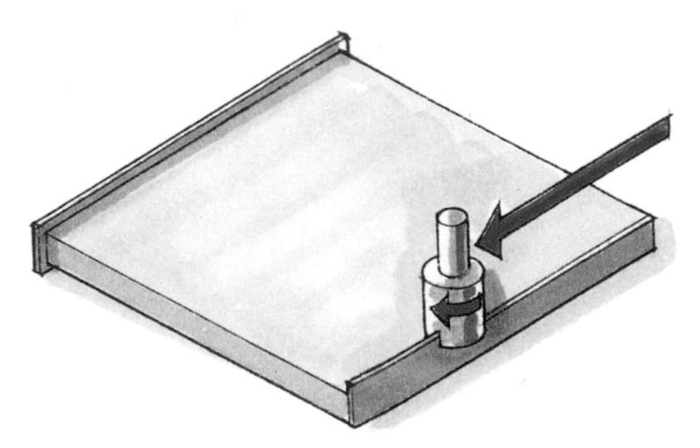

Trimming plastic or wood lippings
Back cutting helps prevent break out when trimming applied lippings flush with the top surface of the base material. Use a cutter with bottom-cutting edges (see page 16) and a stepped fence that lifts the cutter above the vertical overlap of the lipping (see page 13).

FEED RATE

Although cutter-rotation speed can be set fairly precisely, the rate at which the cutter is fed into the wood is entirely dependent on you, the operator. Though it is affected by the hardness or density of the material you are cutting, and the type of cutter you are using, the correct feed rate is more a matter of experience, gained through trial and error.

The main criteria are that the feed rate must not be so slow that friction causes the cutter to overheat, or too fast for waste material to clear quickly.

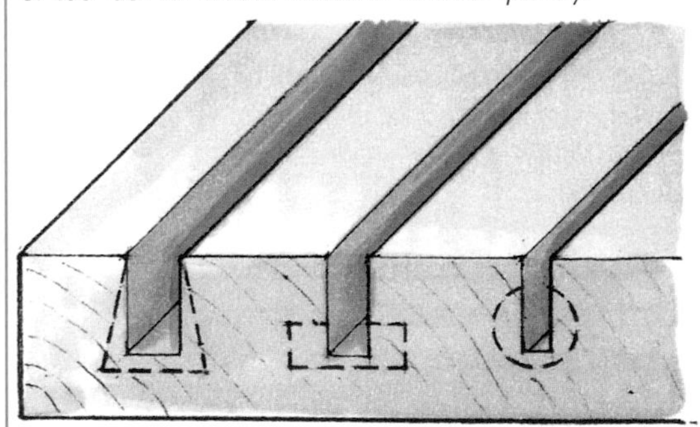

Providing waste clearance
Dovetail cutters, T-slot cutters or ball-slot cutters can only make cuts in a single pass; take extra care when using such cutters, as clearing waste from these 'enclosed' cuts can be a problem. Where the width of the cut permits, pre-cut a narrow slot before working; this reduces the amount of material to be removed.

Routing plastics and aluminium
When routing plastics, set a relatively low feed rate to avoid melting the material, which can weld itself back together behind the cutter or imbed itself in the cutter flutes. Similarly, because it is a soft metal, shards of cut aluminium may pack into the flutes.

CHAPTER 4 During any routing operation, it is essential that a single workpiece or a group of components are held securely, either in a jig or to a rigid work surface or bench. The time spent assembling the necessary cramps, or even making a purpose-made work-board, pays dividends in avoiding injuries and costly damage to the work.

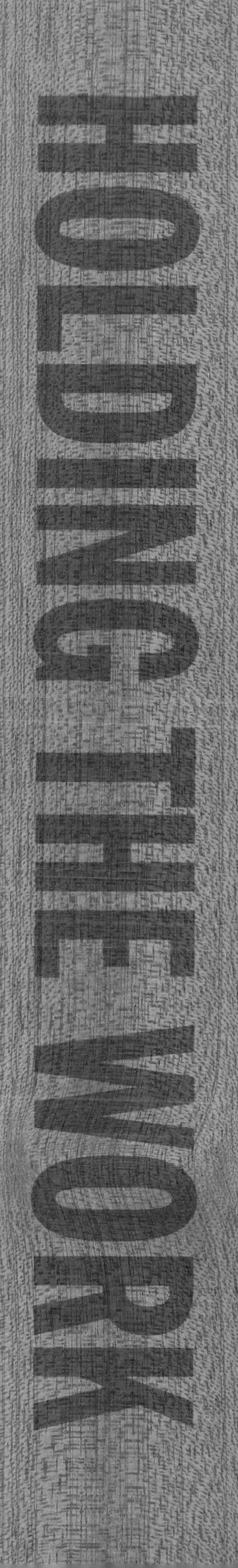

USING VICES AND CRAMPS

Many methods of clamping and holding work can be used for routing; these include G-cramps, sash cramps and bench stops, as well as home-made devices such as folding wedges. Always protect the surface of the wood by inserting strips or blocks of softwood, MDF or plywood between the work and the jaws of metal cramps or vices. Modern sliding-jaw cramps are usually supplied with clip-on plastic facings for this purpose.

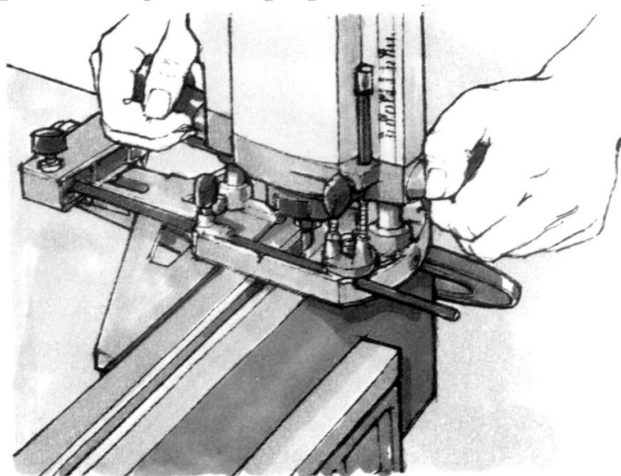

Holding work in a bench vice
Traditional bench vices are only suitable for holding a workpiece vertically, so that the part being machined projects safely above the jaws. Before switching on, always ensure that the cutter will not run into the jaw facings or, worse still, the vice itself. If necessary, fit additional wooden facings or use softening blocks, so that the cutter cannot be damaged accidentally.

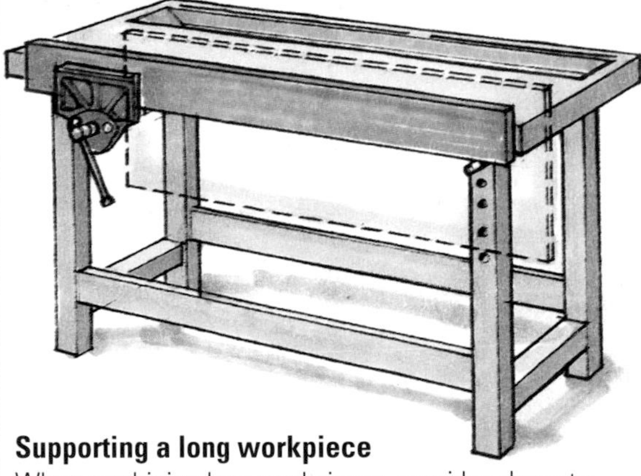

Supporting a long workpiece
When machining long workpieces, provide adequate support at both ends to prevent them pivoting. Clamp the free end to the bench leg, or drill a series of holes down the leg for a removable peg or a holdfast to support the work.

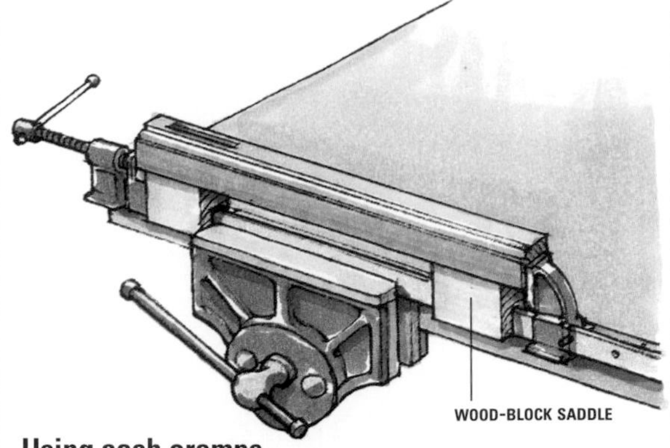

Using sash cramps
Fix a sash cramp in a bench vice to hold a workpiece at both ends. Make a pair of saddles that drop over the bar of the cramp, to prevent the workpiece swivelling.

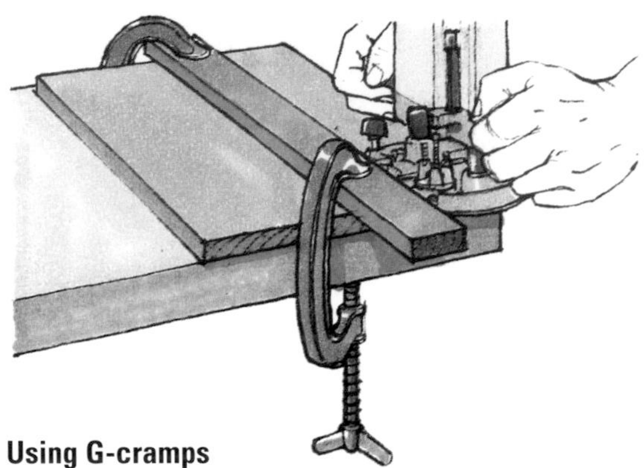

Using G-cramps
For most routing operations, the wood is held flat on a bench or table, so that the router can be fed across the face or along an edge. G-cramps or small fast-action cramps are ideal, provided you place them where they do not interrupt the passage of the router. You can use the cramps to secure the work and hold a guide onto the surface at the same time.

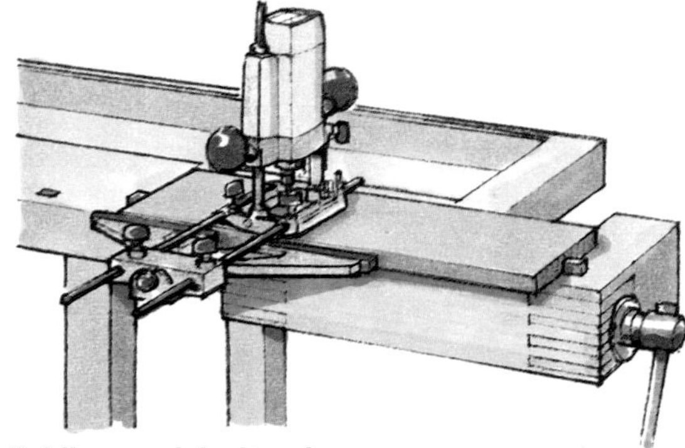

Holding work by its edges
Holding a workpiece by its edges avoids the possibility of cramps projecting above the surface of the wood. One method is to hold the work flat between a pair of metal bench stops on a traditional woodworker's bench; one stop is located in the fixed worktop and the other in the adjustable end vice.

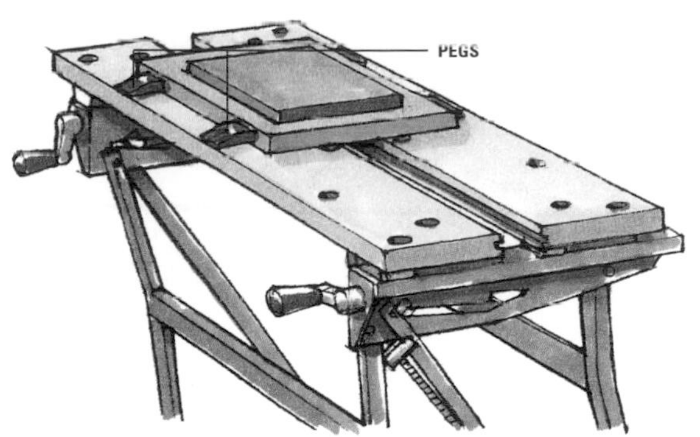

Folding benches
Portable folding workbenches are usually fitted with extended vice jaws, to secure work vertically, and with plastic pegs that hold work flat on the bench top. Each end of the vice opens independently, so you can clamp tapered work between the jaws. A folding bench can support a workboard (see page 38) or jigs, and can also serve as a leg stand for small router tables or stands.

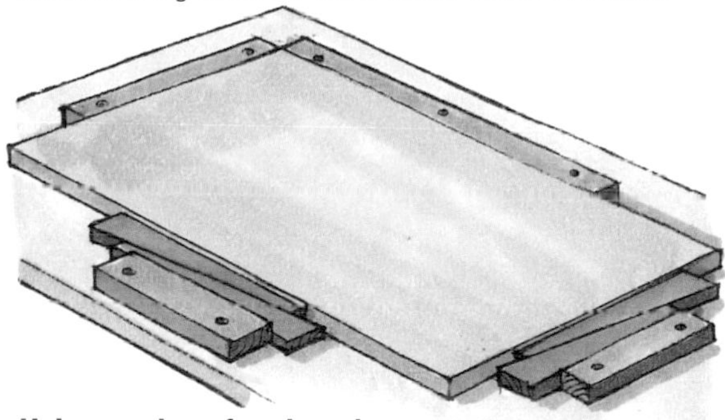

Using wedges for clamping
Nail or screw two battens to a flat board to form a right angle, and butt the corner of the workpiece against them. Fix another batten a short distance from the free end of the workpiece, and drive a pair of 'folding' wedges between the batten and the edge of the work.

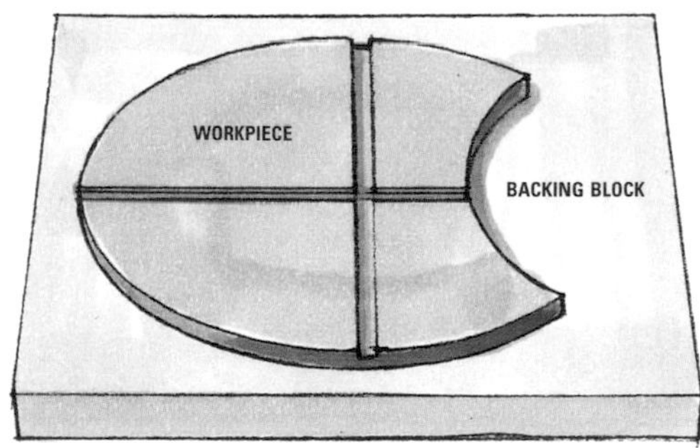

Backing blocks
Fix thin materials or small irregular-shape workpieces to a backing block which can be held firmly. Use a rectangular backing block to provide a straight edge to guide the router's side fence. Fix the work to the block with hot-melt glue, panel pins, PVA adhesive, screws or double-sided tape (see right).

TEMPORARY MEASURES

You don't always need special cramps or vices to hold a workpiece securely. Use fastenings such as wood-screws, or even panel pins, to pin the wood onto a flat board or workbench, but make sure these fixings are positioned so that they will not damage the router or spoil the workpiece. Hot-melt adhesives are a good alternative to metal fastenings, but apply glue only to waste areas of the work.

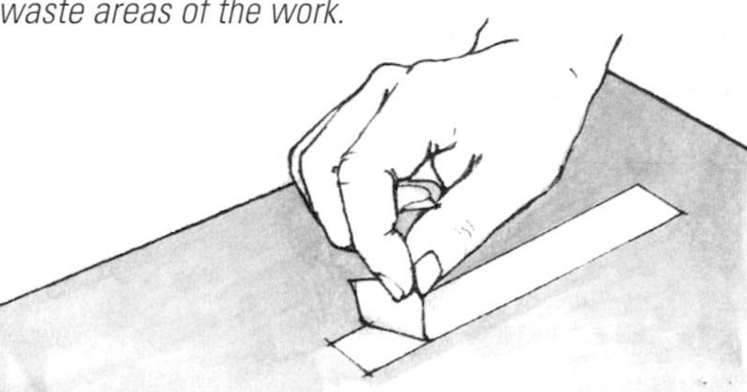

Using double-sided tape
Double-sided adhesive tape can be used for temporary holding. Two types are available, a thin stationery tape and a thicker, heavy-duty carpet grade; both work best between smooth, flat surfaces, free from dust and grease. Seal the surface of the workboard and, if possible, the workpiece, with sanding sealer or light varnish, or face the workboard with plastic laminate.

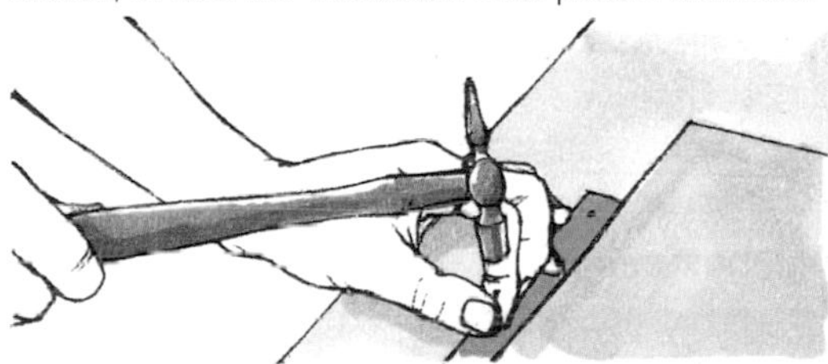

Additional support for tape
When there is only a small area of contact between the wood and worktable, double-sided tape can 'creep' or move very slightly sideways, especially the thicker grades. For additional support, pin narrow battens alongside the workpiece.

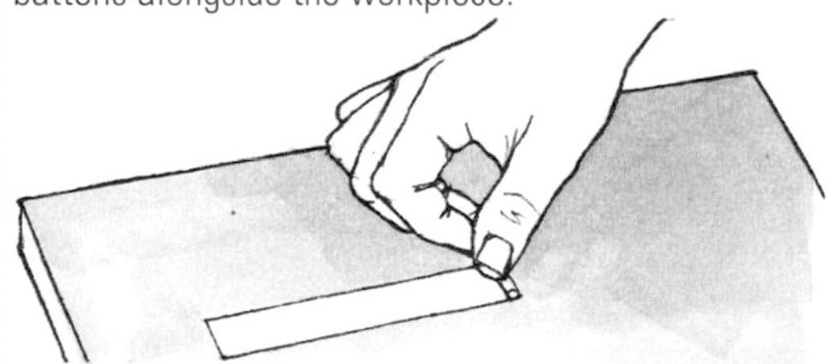

Removing tape
Double-sided tape rarely requires a solvent; roll it off the wood with your thumb. Test the tape beforehand on a piece of scrap material, to make sure it does not lift surface veneers or laminates.

MAKING A WORKBOARD

When working both wide and narrow materials, it is best to keep the work fully supported by laying it on a flat bench. Lay particularly large pieces over a temporary top of thick sheet material supported on trestles, or even on a flat floor covered with a sheet of 3mm (⅛in) hardboard or plywood. This allows the cutter to be set slightly deeper than the thickness of the workpiece where required. An even better method is to make a special-purpose workboard, incorporating cramps and stops. It can be designed to hold any shape of panel or workpiece, and can be used to clamp a number of similar components for simultaneous machining.

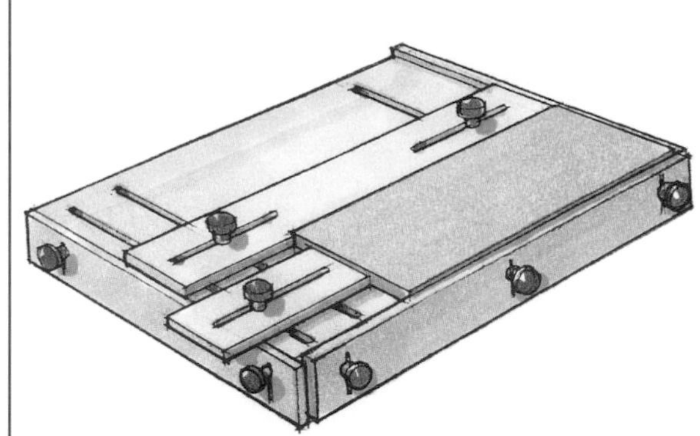

Gripping the work
Butt the work against the end and side stops, and adjust the sliding stops to hold it securely on all sides.

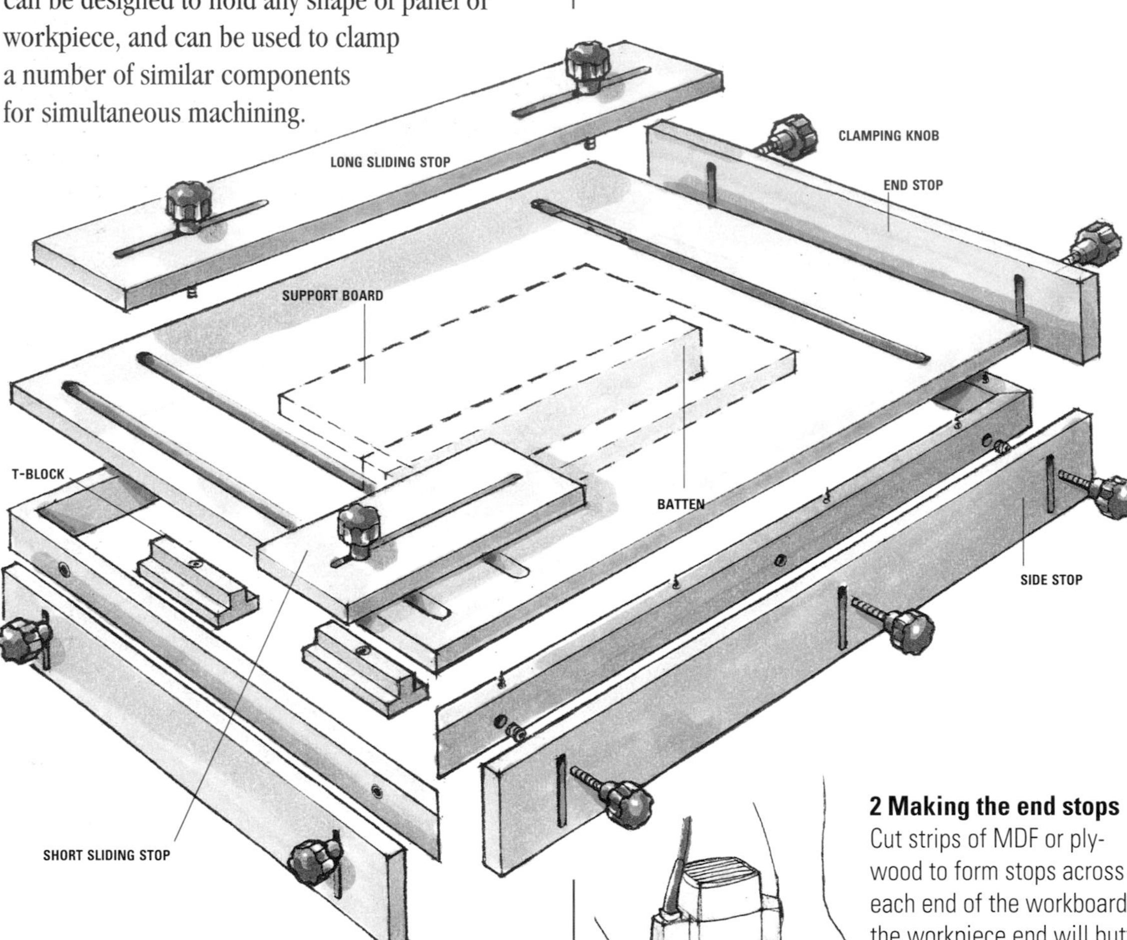

1 Cutting the board to size
Cut the workboard from 12 to 18mm (½ to ¾in) thick MDF or plywood. An overall size of 600 x 900mm (2 x 3ft) is suitable for a wide range of furniture projects, but you can vary the board size to suit your own needs. Machine or plane square edges on the board, and check that the sides are parallel and the corners perfectly square.

2 Making the end stops
Cut strips of MDF or plywood to form stops across each end of the workboard; the workpiece end will butt against them. You can screw the stop to the edge of the board, but it is better to be able to adjust the height. Cut an adjustment slot 75mm (3in) from each end of the stops, using a straight cutter.

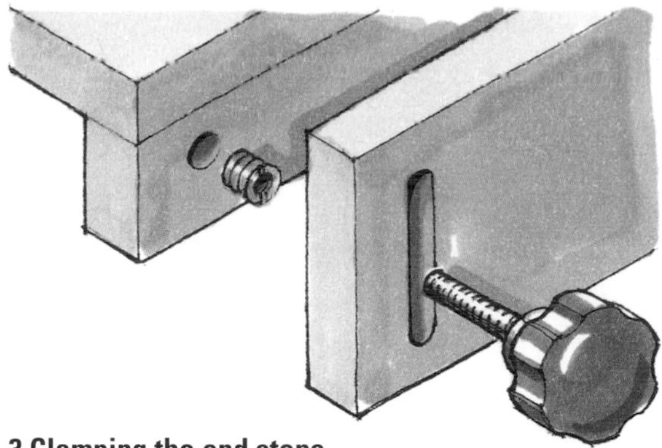

3 Clamping the end stops
Screw four battens to the underside of the workboard, flush with the edges. To hold the adjustable end stops in place, screw a pair of threaded inserts into the end battens, to align with the slots cut in the end stops. Pass the threaded shaft of a clamping knob or lever handle through each slot into the captive inserts. These fittings are available from router-accessory suppliers.

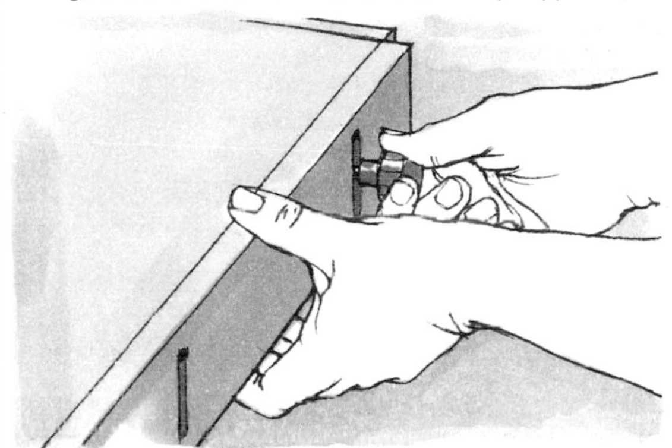

4 Making the side stop
Make a similar adjustable stop along one side, and clamp it to the front with three clamping knobs.

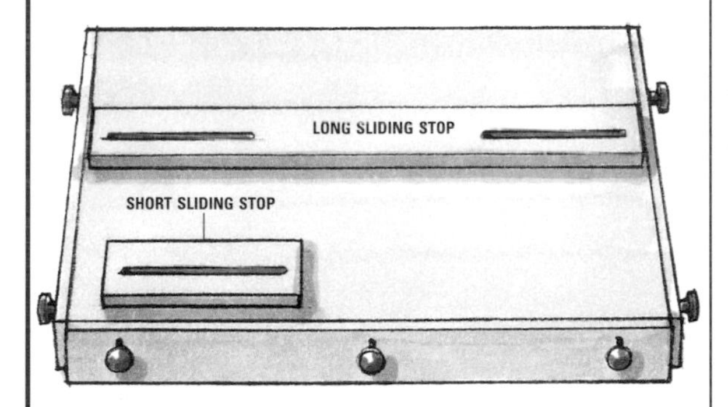

5 Making the sliding stops
Make top-mounted long and short sliding stops, to hold the workpiece against the end and side stops. The longer side stop has a pair of slots machined longitudinally; a single slot is machined in the shorter end stop.

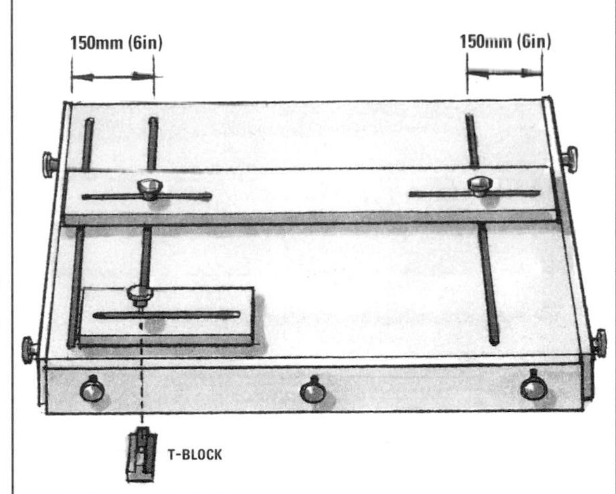

6 Fitting the sliding stops
To secure the sliding stops to the board, make T-blocks that slide in slots cut through the board from front to back. Place one slot about 150mm (6in) from each end of the board; the third slot, cut close to one end of the board, is to hold long workpieces. Fit threaded inserts in the T-blocks for clamping knobs or lever handles.

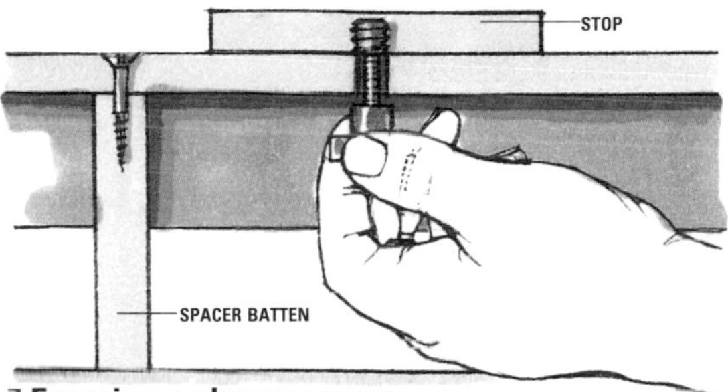

7 Ensuring a clear passage
To make sure the knobs or levers cannot impede the passage of the router, you can mount the board on spacer battens to raise it off the bench; fit the knobs below the raised board, with the inserts screwed into the stops.

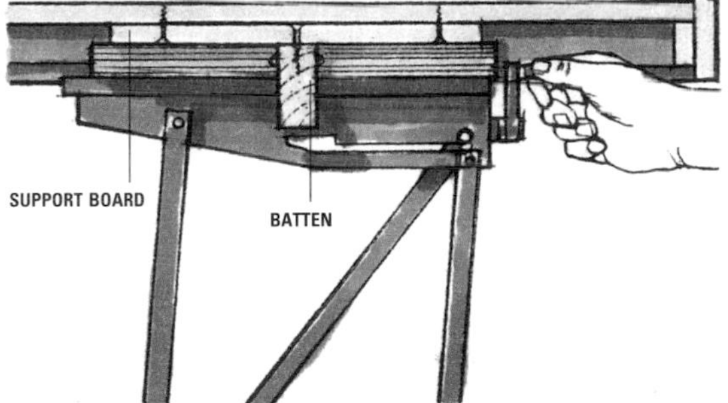

8 Fitting to a folding bench
To mount the workboard on a folding bench, screw a stout batten to a support board that is then screwed beneath the workboard. Clamp the batten between the vice jaws. The support board provides stiffness and room to operate the bench's vice-jaw controls.

FITTING CRAMPS TO THE WORKBOARD

Gripping the edges of a workpiece between stops is usually enough to hold it firmly to the board. However, it pays to incorporate additional cramps for when you want to hold workpieces down onto the board, or to secure guide fences. Machine a matrix of T-slots in the workboard, for special toggle cramps to hold work of any shape or size.

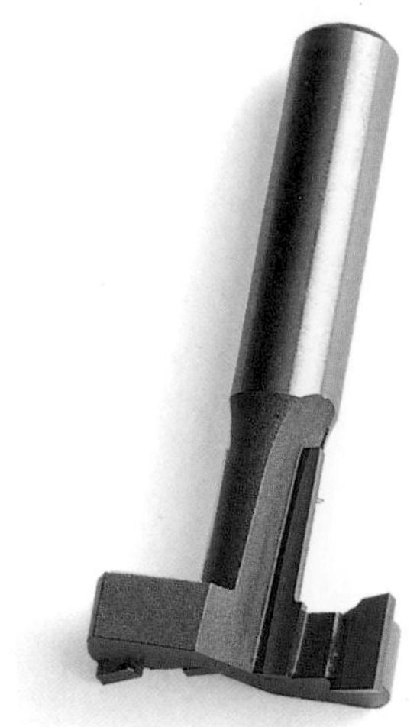

T-slot cutters
These cutters, originally designed for producing demountable display-shelving and racking systems, can be used to cut the matrix of slots. Guide the router using a straightedge or T-square clamped across the board (see pages 47–8).

Machining the board
Cut the basic workboard from 32mm (1¼in) MDF. Router the matrix first, using a straight cutter, then router out the base of each slot with the T-slot cutter.

Machine a matrix of slots across a workboard

100mm (4in)

100mm (4in)

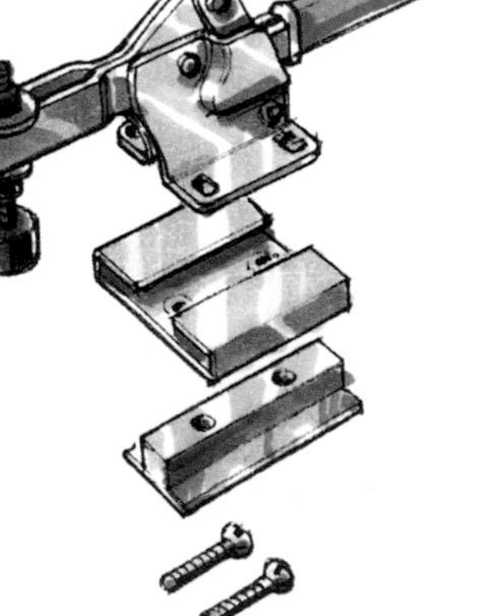

Toggle cramps and mounting plates
The base of the toggle cramp is located in a bent-metal mounting plate. This plate is bolted to a T-section nut that slides in slots cut across the workboard.

USING WORKSHOP CRAMPS

You can use ordinary fast-action cramps or G-cramps to hold the work onto a workboard or bench. To leave the top surface free from obstructions, use an end-socket cramp that fits in a hole drilled in the end of the workpiece. The same cramp can be used to secure guide fences or to clamp a workpiece beneath a stout batten running across the width of the workboard.

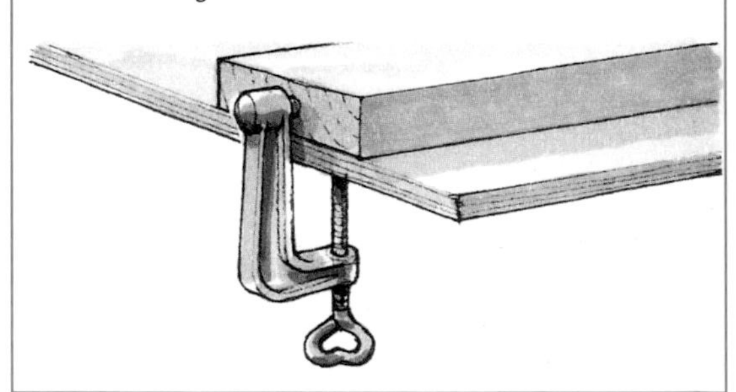

VACUUM CLAMPING

Vacuum clamping is generally used in professional workshops for fast, repetitive batch work. It can prove expensive for a home craftsperson, due to the high cost of the vacuum pump; however, vacuum chucks for holding specific projects can be easily and cheaply made. Another alternative is to use a domestic or industrial vacuum cleaner; check beforehand that the unit is suitable, since continuous airflow restriction may damage it.

Materials and equipment

You can obtain everything you need to make a vacuum cramp from router-accessory suppliers, including a small professional-standard pump unit equipped with a foot-operated on/off switch. Neoprene sealing strip is available as a flat self-adhesive material or as a round strip, for setting into a routed groove. Neoprene can also be obtained in sheet form.

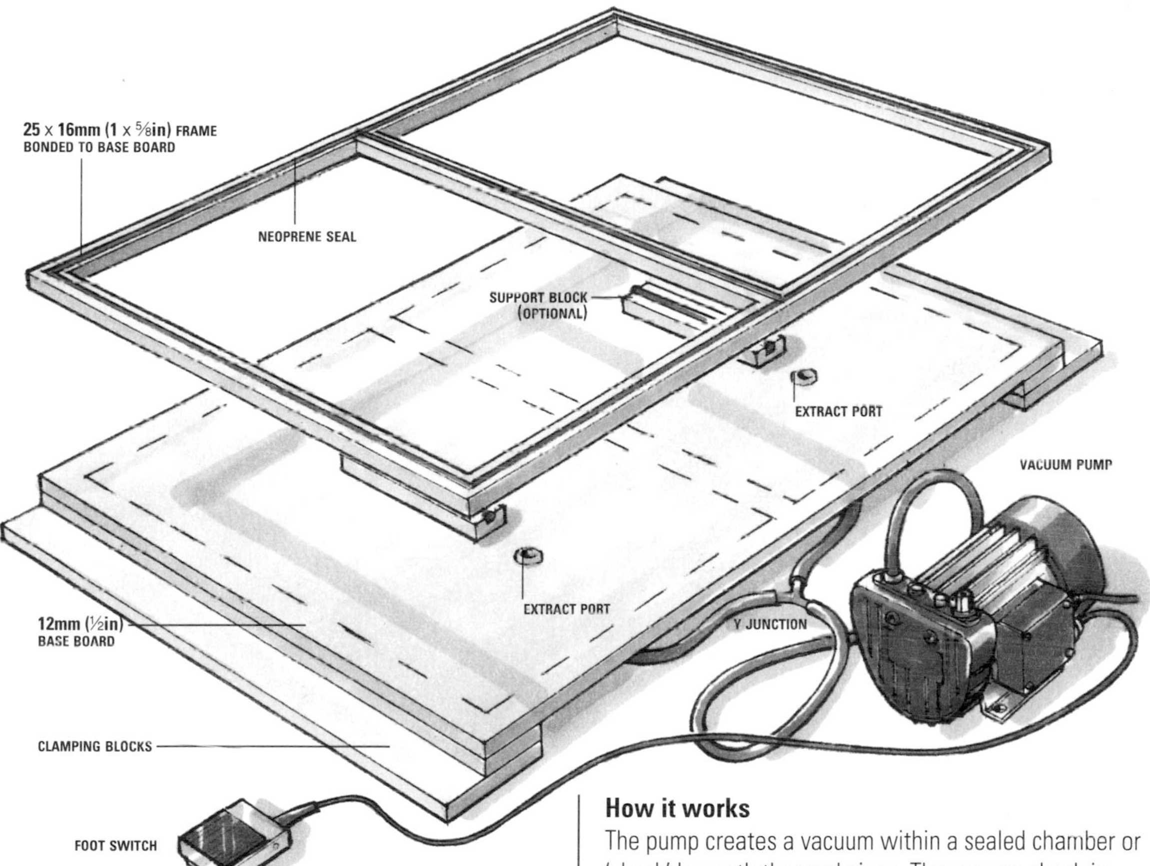

Dimensions

The proportions of the chuck or chamber should be as close as possible to the size of the workpiece. For most projects, a vacuum chuck of 600 x 900mm (2 x 3ft) is satisfactory; this can be divided into two usable chambers. The depth is dictated by the frame and neoprene strip.

How it works

The pump creates a vacuum within a sealed chamber or 'chuck' beneath the workpiece. The vacuum chuck is made using a compressible neoprene sealing strip set into or stuck to a flat surface. An extract port drilled in the chamber is connected to the pump, allowing air to be withdrawn to form a vacuum beneath the workpiece; external air pressure holds the work firmly in place. By forming two or more chambers, the chuck is reduced in size by plugging the extract ports in the areas not covered by the work.

Applying sealing strip

The chamber need not be rectangular, as the seal groove can be cut to follow any shape. Chamfer the edges of the groove so that the strip deforms to make a tight seal when compressed.

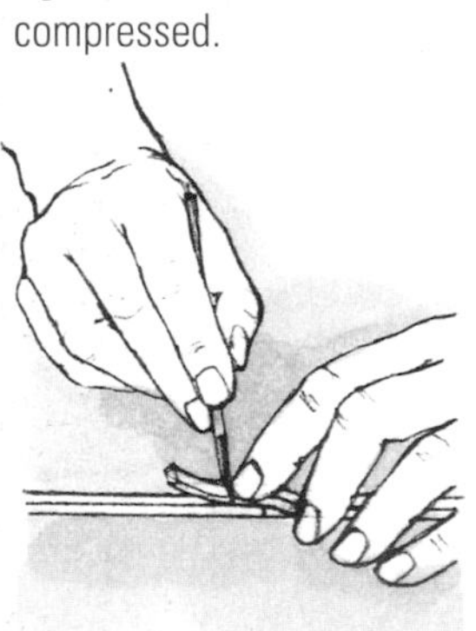

Using self-adhesive strips

For small chucks, apply self-adhesive strips directly to the base board, following the shape of the work. Ensure that surfaces are dust- and grease-free, and seal porous surfaces with sanding sealer or varnish. Overlap end-to-end joints, and cut through both thicknesses with a knife.

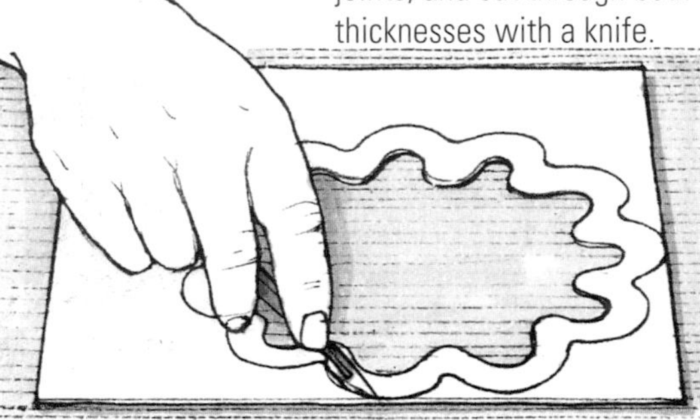

Applying self-adhesive sheet

For intricate shapes, trace or draw the shape onto the surface of the neoprene sheet, and cut it out before applying it to the baseboard.

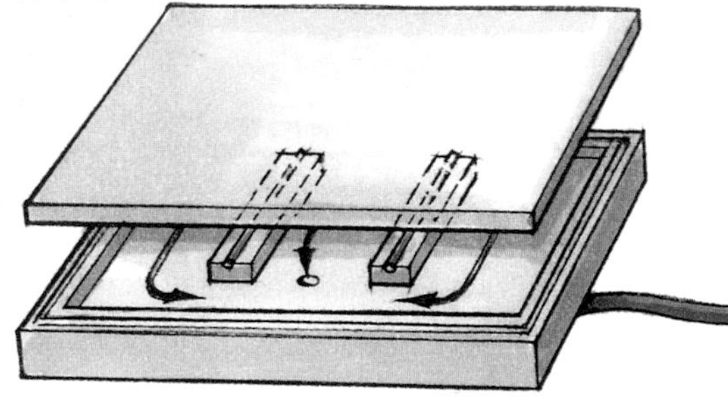

Supporting the workpiece

To prevent a thin flexible workpiece bowing when the vacuum pump is switched on, support it at regular points across the surface of the chuck. Glue thin blocks or strips of wood within the vacuum area, making sure they lie slightly below the neoprene strips, to give an adequate seal around the edges of the chuck.

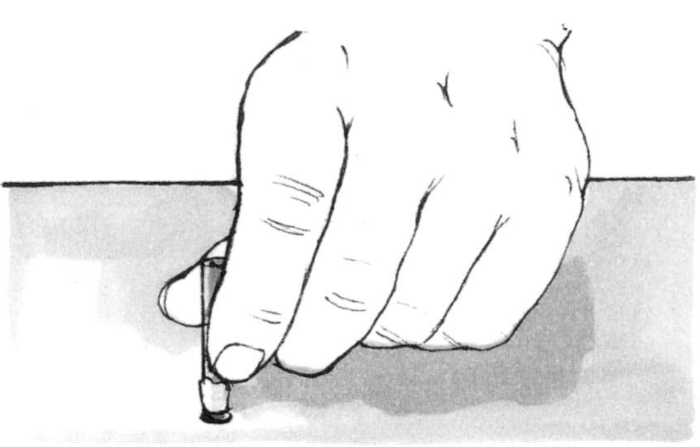

Inserting an extract port

Drill a hole in any convenient position in the vacuum chamber to take a short length of 3 to 4mm (⅛ to 5⁄32in) diameter brass or aluminium tube. Glue the tube in the hole, using an epoxy adhesive. Connect a length of plastic hose between the pump and tube.

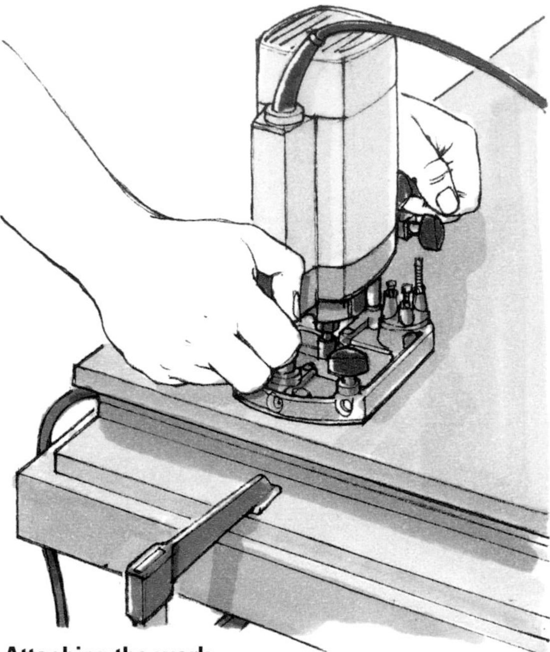

Attaching the work

Attach the chuck securely to a rigid bench, and connect the pump or vacuum extractor by the extractor hose. Place the work over the chuck, checking that it covers the seals along each edge to form the vacuum chamber. Switch on the pump before routing; to release the workpiece, switch the pump off. Do not leave a pump or extractor unit running longer than necessary.

Although the workpiece can overhang the sides of the chuck, avoid excessive overhang, as the weight of the router can lift the workpiece away from the seals.

Where work needs to be turned, when rabbeting or moulding more than one edge, or repeating a cut on both faces, secure the workpiece over the chuck and lay over it a template fitted with locating blocks. The weight of the router keeps the template in position.

Chapter 5

The inherent precision of the portable router means very little without the wherewithal to guide the cutter on its intended path. Most routers are packaged with a fence and a guide bush or two, but it is the number of accessories and templates that you can buy or make yourself that elevates the router above the status of a power-operated plough plane.

WORKING WITH SIDE FENCES

When making straight cuts, either parallel to or along the edge of the work, use the side or parallel fence supplied with the router.

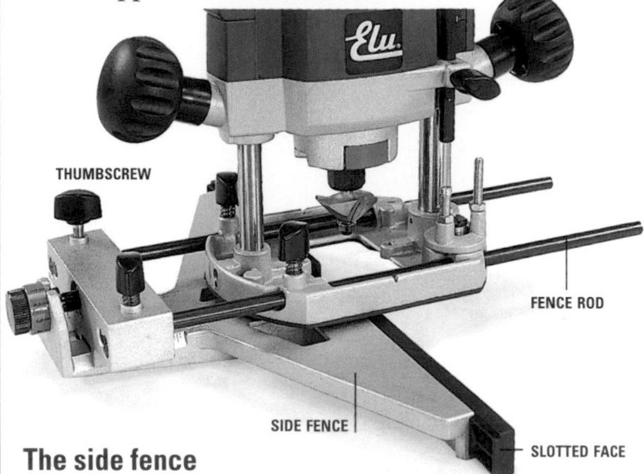

The side fence

A typical side fence is supplied with a pair of independently adjustable plastic faces, or is drilled to fit wooden faces. Fixing screws are located in slotted holes, with their heads recessed below the surface of each face. The slotted screws allow for sideways adjustment, so that the inner end of each face can be set close to the cutter mounted between them, preventing the fence pulling into the work at the start and finish of the cut and damaging the wood with the cutter. There must, however, be sufficient clearance to prevent waste packing around the cutter.

The fence is attached to metal rods, mounted on the router base plate. The distance of a groove or housing from the edge of a workpiece is limited by the length of the fence rods supplied with the router. Longer fence rods should be stiff enough to prevent excessive flexing; when using them, fit an extended one-piece face to the side fence (see bottom right).

AVOIDING OBSTRUCTIONS

Take care that the protruding ends of side-fence rods or extended fence faces cannot catch on cramps or other obstructions while routing. Also make sure that the power cable cannot become trapped between the guide edge and fence.

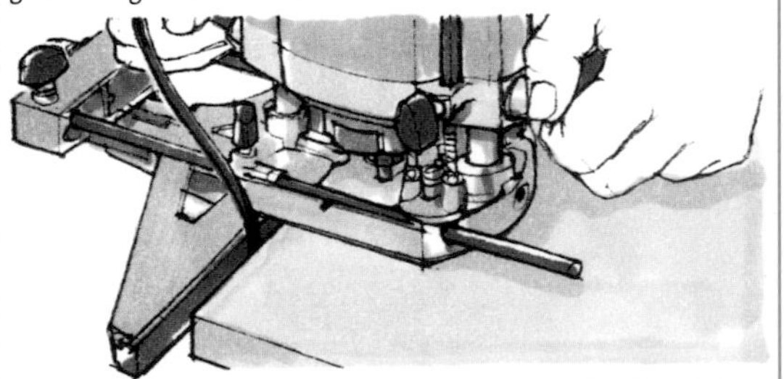

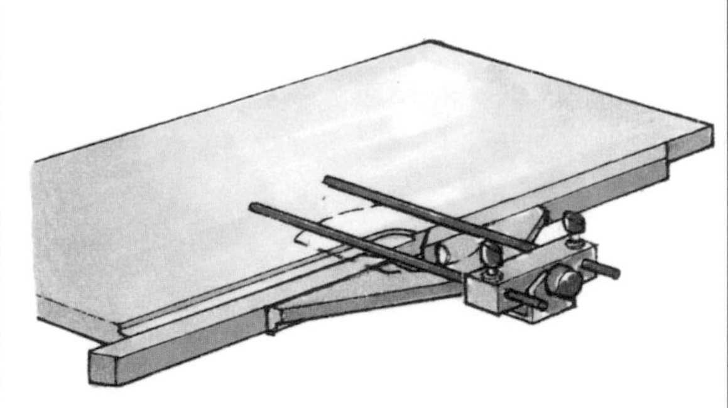

Fitting longer faces

Screwing longer faces to the fence reduces the risk of it pivoting and turning the cutter into the work at each end of the cut. Longer faces also allow the fence to be held more firmly against the guide edge throughout a routing operation.

Using end battens

An alternative method of preventing the cutter from turning in at the start and finish of the pass is to clamp waste battens at each end of the work. This also eliminates break out at the end of the pass.

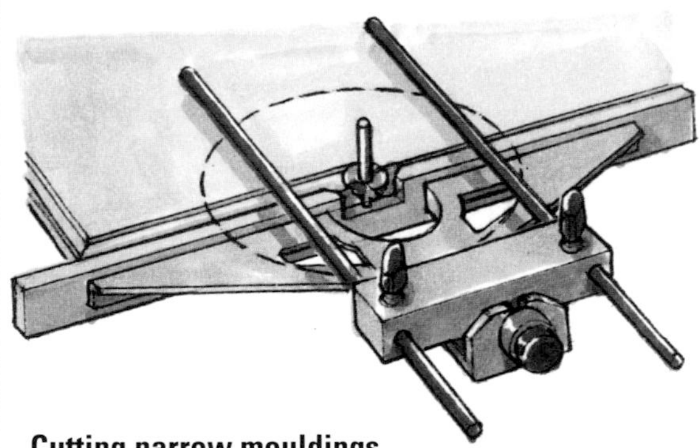

Cutting narrow mouldings

When cutting a narrow moulding – one that does not involve removing the entire edge of a workpiece – a deeper face can be fitted and the cutter plunged only partway through it. This provides, in effect, a continuous fence, eliminating the risk of the inner ends of conventional faces catching on the work, and reduces the possibility of the fence turning in at the beginning and end of a cut.

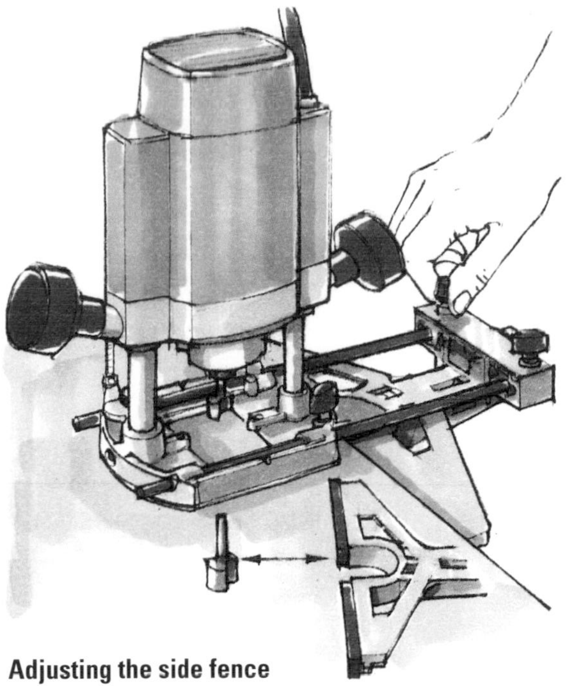

Adjusting the side fence

To place a cutter the required distance from the edge of the work, slide the fence along the mounting rods and secure it, using the integral cramps on both the fence and the router's base plate. For grooves or panel mouldings, set the fence by measuring the required distance from its face to the tip of one flute of the cutter. On some models, precise adjustments can be made using the side-fence microadjuster.

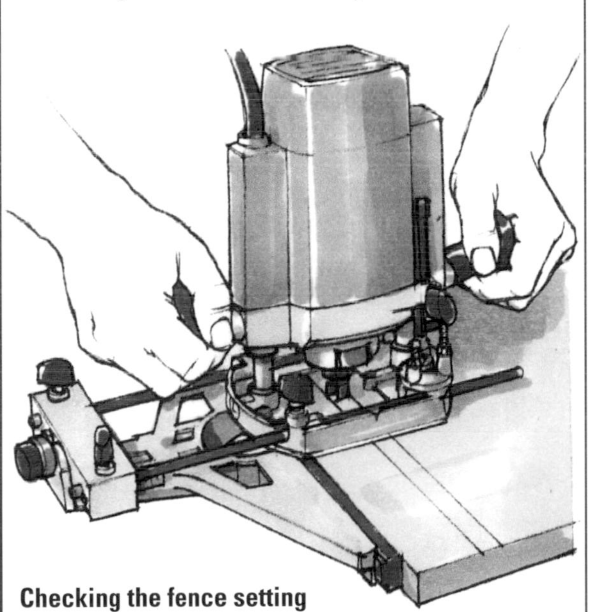

Checking the fence setting

Mark out the position and width of the required cut on the face of the workpiece. Check the cutter and fence positions by placing the router over the work, with the cutter resting on the surface of the wood. When cutting several identical workpieces, you need only mark out and check the router settings against one of them.

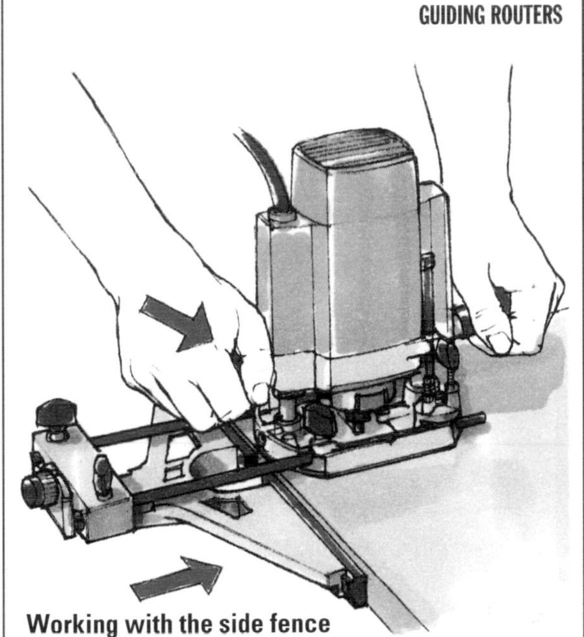

Working with the side fence

Set the required depth of cut with the router's depth stop, and set its turret stop for making a deep cut in stages (see page 12). For any side-fence operation, the edge of the work (the guide edge) must be finished straight; remove any high points or deposits of resin or glue. While making the cut, keep the fence faces held tight against the guide edge.

When you are moulding an edge, it does not matter if the fence happens to move away from the guide edge, since a second pass will rectify the error. However, any sideways movement while grooving or panel moulding will deflect the cutter from what should be a straight cut, parallel with the edge.

FEED DIRECTION

For most side-fence operations, the feed direction should oppose cutter rotation. If necessary, light back-cutting techniques (see page 34) can be used for edge-trimming or moulding operations, but should never be used when cutting grooves or panel mouldings, as the rotational direction of the cutter tends to pull it off line, forcing the fence away from the guide edge.

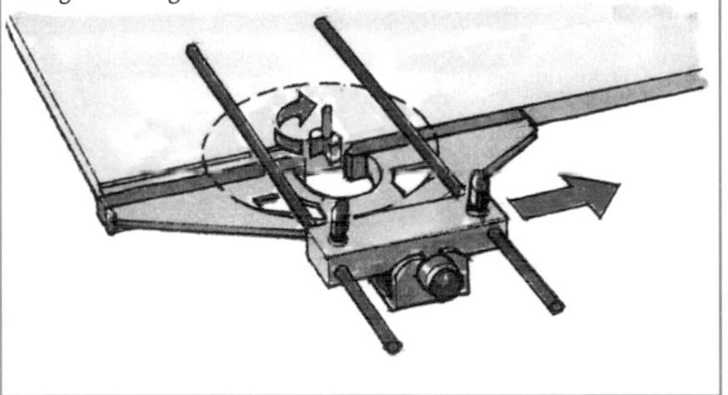

AUGMENTING THE GUIDE EDGE

Always ensure that the guide edge is wide enough to accommodate the router fence.

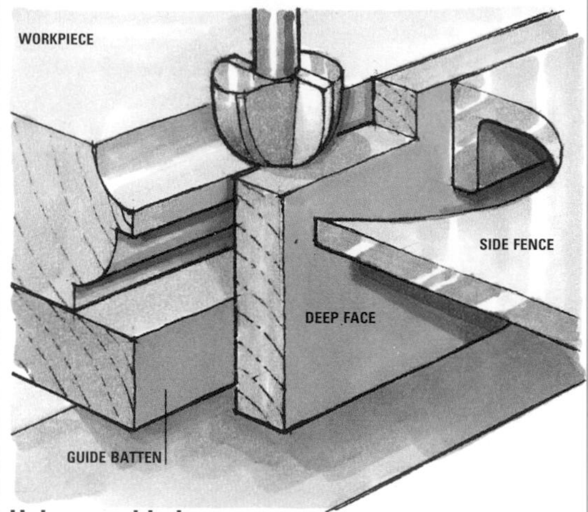

Using a guide batten

If cutting a moulding leaves you with a very narrow square edge, clamp a guide batten beneath the workpiece. Alternatively, the workpiece can be clamped flush with the edge of the workbench and the fence run against that. If necessary, screw deeper faces to the fence.

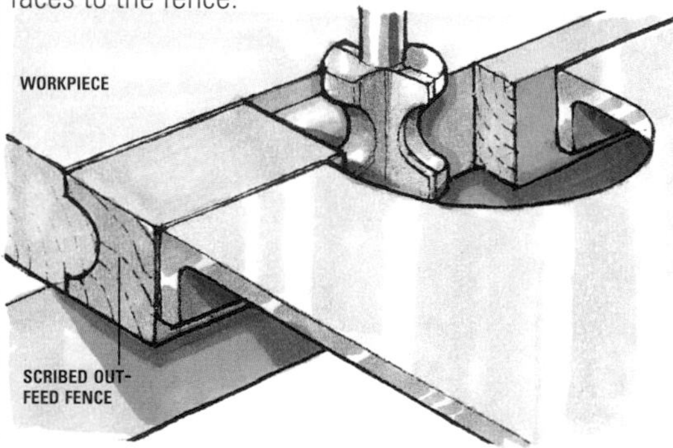

Fitting a scribed out-feed fence

When using edge-moulding cutters that remove wood across the full width of the guide edge, replace the rear face of the fence with one that is scribed to the reverse profile of the cutter. This has the effect of providing a bearing surface behind the router. Two alternative rear-face profiles are shown below.

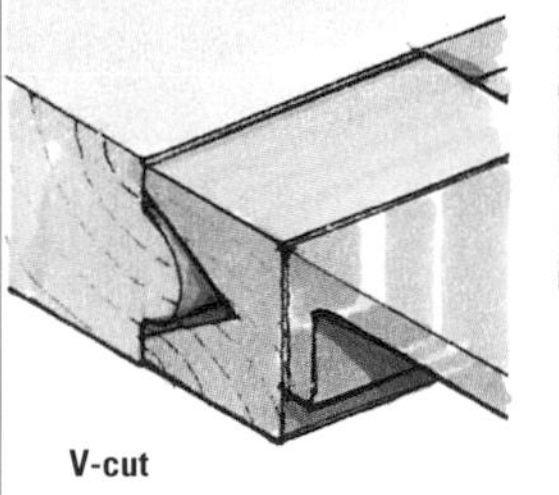
V-cut

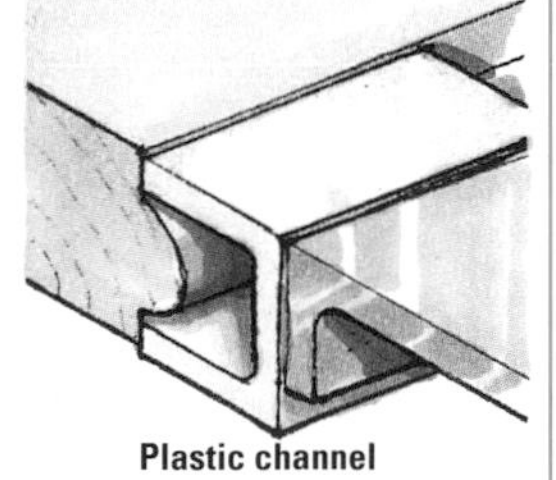
Plastic channel

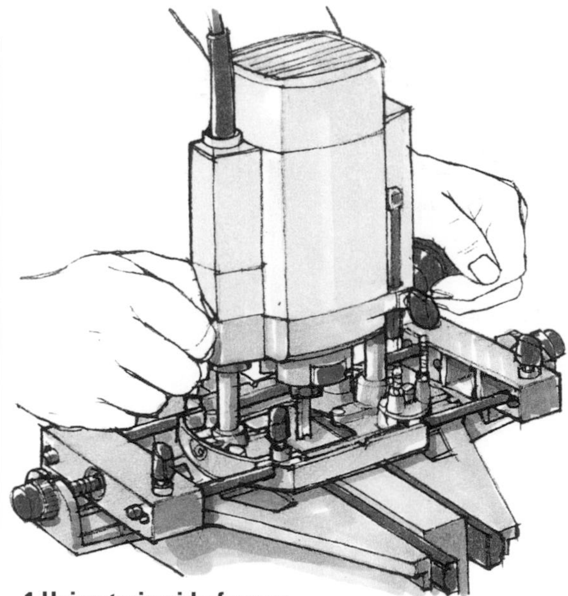

1 Using twin side fences

When cutting mortises, rabbets or mouldings on narrow workpieces, use two side fences, one mounted on each side of the router, running against both faces of the work. You may need to fit extra-long side-fence rods.

Check that the faces of the workpiece are parallel, as any tapering will cause the fences to jam or wander from side to side. Set the distance between the fence faces to leave minimal side play, but enough to keep a steady feed at a constant rate. When working with twin side fences, the cutter can be fed in either direction.

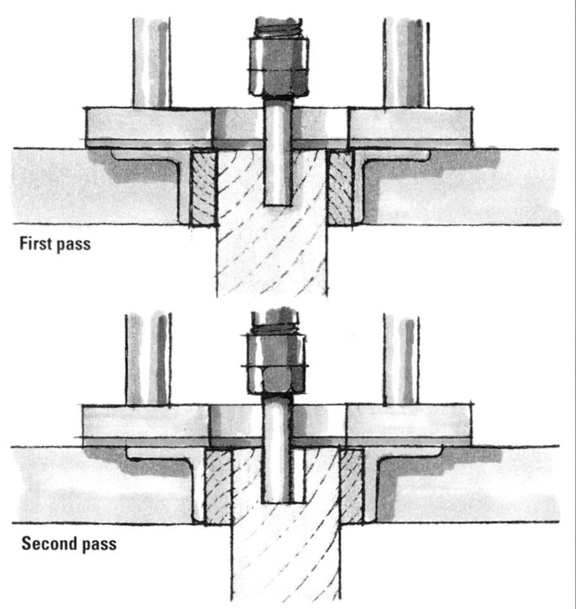

2 Centring the cut

To centre a cut precisely when using twin fences, make the first pass, then turn the router round and make a second pass in the opposite direction. Check the finished width of the groove, and set the thickness of the tongue on the tenon to match.

STRAIGHTEDGE GUIDES

Housings, slots and other surface cuts can be made easily by running a portable router against a straightedge clamped square or at an angle across the face of a workpiece. For this purpose, most routers have at least one straight side to the base plate, or have a circular base that is concentric to the centre of the cutter.

For accuracy and safety, ensure that a straight-edge guide is rigid enough not to bend or flex, and always clamp it securely to the work. Check that cramps or other fixings do not impede the path of the router. Use a straightedge that is longer than the width or length of the workpiece, to allow the router to be fed square into and out of the cut.

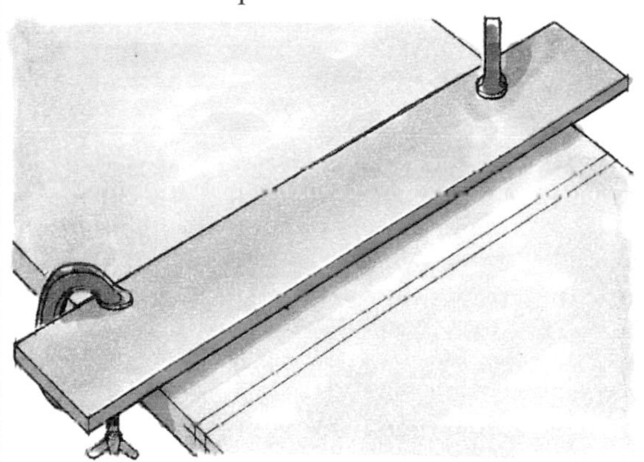

Making a straightedge
Cut a straightedge from stable material such as MDF, plywood, polycarbonate or other shatterproof plastic. Cut the straightedge at least 150mm (6in) wide and 6 to 9mm (¼in to ⅜in) thick. Plane at least one guide edge straight and square.

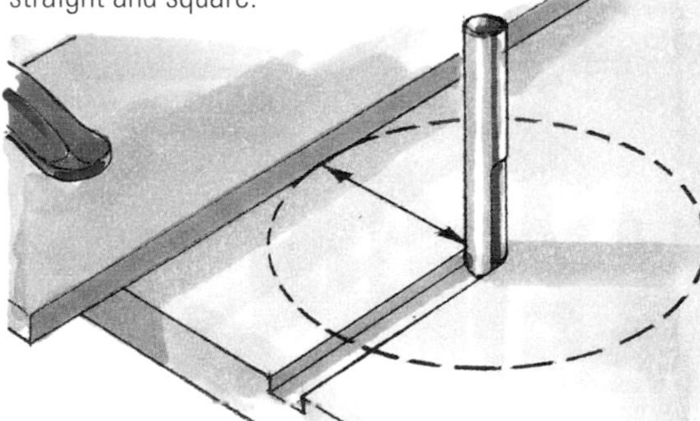

Positioning a straightedge
Mark out a housing, for example, on the surface of the workpiece. Measure the distance from the edge of the router base to the cutting edge of the cutter. Clamp the straightedge an equal distance from the housing.

Base-plate facings

Cutting regularly spaced housings
A spacer batten screwed to the underside of a straightedge allows you to cut housings at regular intervals along a workpiece. The batten must fit the housings exactly. The first housing is cut parallel to the end of the board, using a side fence or another straight-edge. The spacer batten is then dropped into each subsequent housing, to position the guide for cutting.

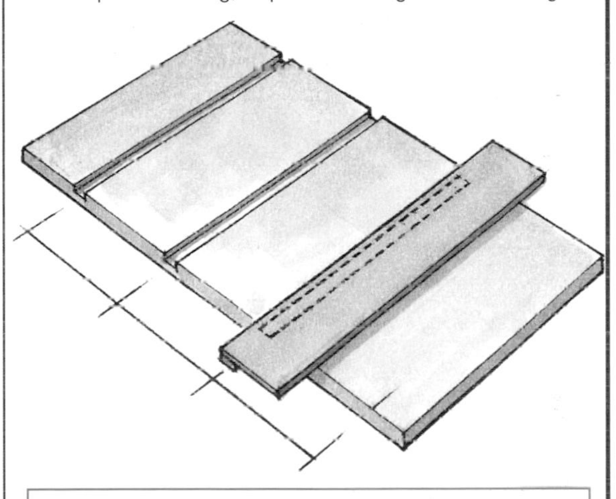

FEED DIRECTION

When working against a straightedge, feed the router from left to right; the rotation of the cutter pulls the router into the guide edge rather than allowing it to wander away.

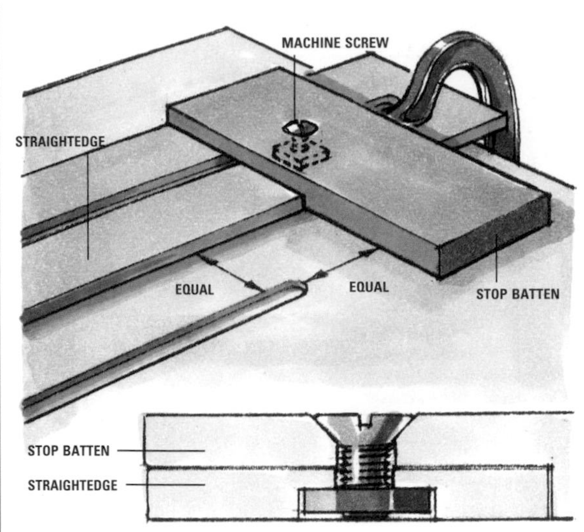

Cutting stopped housings
You can raise the cutter when you have run the router up to a pencil mark on the straightedge, but for greater accuracy, screw short stop battens to the straightedge to limit the router travel. Adjustable stops can be fitted, using countersunk machine screws set into threaded metal plates that run in a groove cut along the underside of the straightedge.

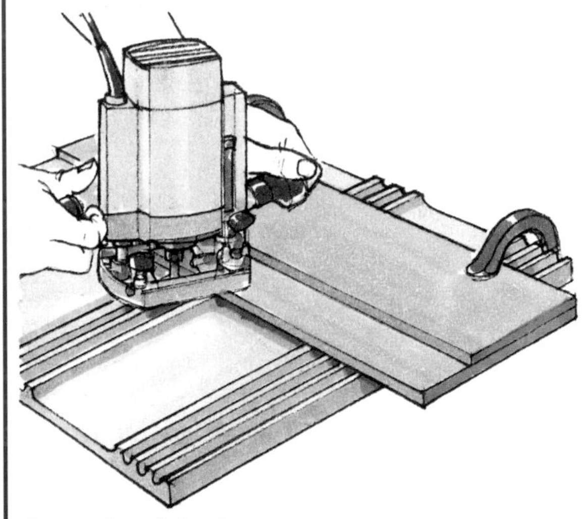

Stepped straightedges
To cut across an uneven surface, use a stepped straightedge made up from two strips of sheet material. Cut one strip 63mm (2½in) wider than the other, and glue them together with their rear edges flush. Fit a straight cutter in the router and, with the base running against the edge of the upper strip, trim the wider strip parallel to it. Attach a stock if you plan to make a number of cuts.

Clamp the stepped straightedge across the work as before, but increase the router's cutting depth by the thickness of the bottom strip. Pass the router across the stepped face, using the upper strip as the guide edge.

MAKING A T-SQUARE

To align a straightedge square to the edge of a workpiece, attach a stock to form a T-square.

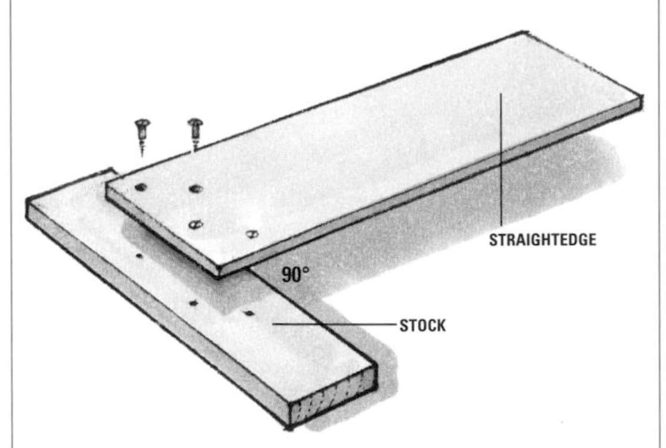

1 Attaching the stock
Cut a 400mm (1ft 4in) long stock from 75 x 12mm (3 x ½in) hardwood. Glue and screw the stock across the end of the straightedge blade, checking that they form a perfect 90-degree angle.

2 Cutting the centre guide
Clamp the T-square to a piece of waste board, with the stock tight against its edge. Fit a straight cutter in the router and set the cutting depth to one-third the stock's thickness. With the router running against the T-square blade, machine a slot across the stock. Next, fit a V-groove cutter, and run the router against the T-square to cut a fine line down the centre of the slot.

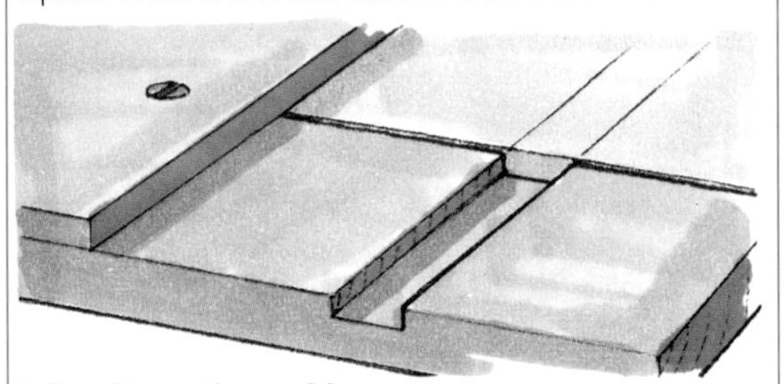

3 An alternative guide
As an alternative, cut the slot in the stock with the same cutter you intend to use for the housing, and align the slot with the housing shoulders marked on the work. If you want to use the square with a different-size cutter later, glue a hardwood block into the existing slot, then cut a new one.

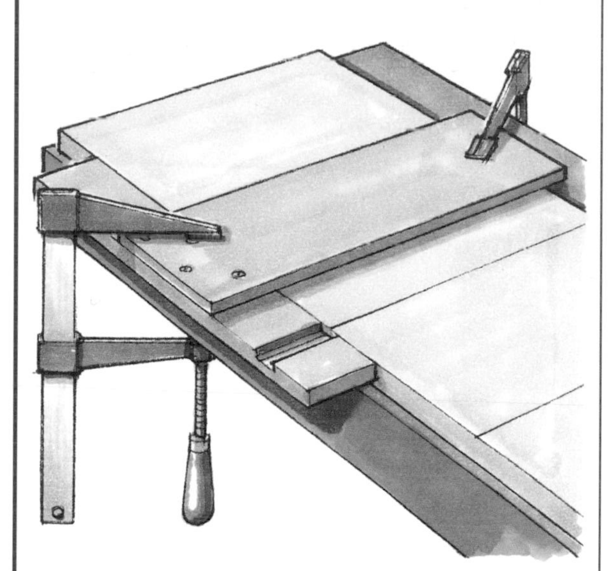

Positioning a T-square

To position a T-square accurately, align the V-groove in the stock with the centre line of a housing, for example, marked on the edge of the workpiece.

Making angled T-squares

To cut housings at an angle across a workpiece, attach the stock to the blade with a coach bolt and wing nut. Cut a notch in the stock to make room for a G-cramp. For a few cuts only, you can screw a stock to a straight-edge at the required angle.

ROUTING FRAMES

For one wide, or two parallel, housings, run the router between two linked straightedges. An adjustable frame, using coach bolts and wing nuts, allows you to set the straightedges to cut housings at any angle. Use a similar frame with intermediate stop battens to cut panel-face mouldings or recess pockets in wood.

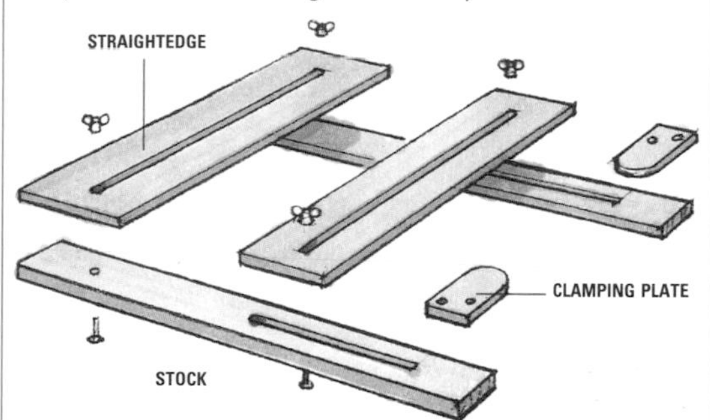

1 Making a routing frame

Cut two stocks 600 mm (2ft) long from 75 x 12mm (3 x ½in) hardwood. Drill a single 6mm (¼in) hole at one end of each stock, and cut 6mm (¼in) wide central slots along the remaining length. Cut similar grooves in a pair of 150mm (6in) wide straightedges. Cut a guide slot across both stocks (see opposite).

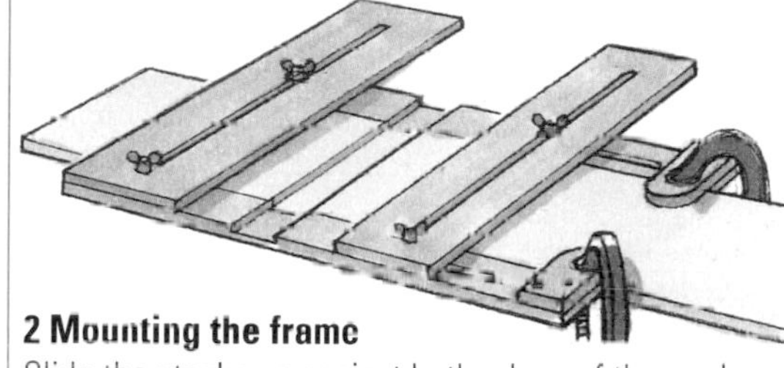

2 Mounting the frame

Slide the stocks up against both edges of the work-piece. Set one straightedge to the required angle and tighten its wing nuts. Adjust and bolt the second straightedge, check the positions and angles, then clamp the frame down, using wooden plates screwed to the stocks.

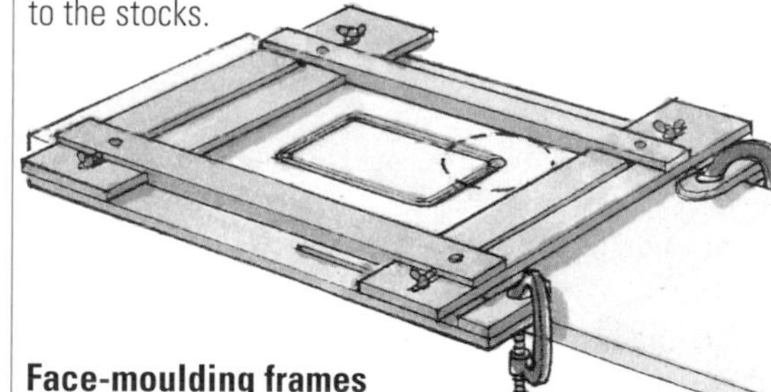

Face-moulding frames

To mould the face of a square or rectangle, fix two battens between the straightedges of a routing frame. Screw these in place, or fit them with machine screws set into threaded plates (see top left, opposite). To cut on the inside of a frame, feed the router clockwise.

USING GUIDE BUSHES

Guide bushes offer a very simple and direct method of guiding the router. They can be used in conjunction with templates, straightedges or, for many applications, the edge of the workpiece itself. A guide bush is similar to a metal ring fence, standing proud of the underside of the router's base plate and concentric to the cutter. This ensures that a constant distance is maintained between the tip of the cutter and the template or guide edge throughout a routing operation.

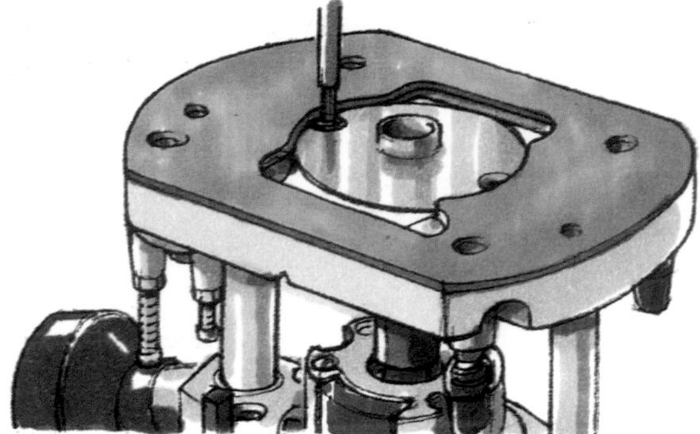

Mounting guide bushes
Router manufacturers use various methods for mounting guide bushes; the most common is to locate the bush with two countersunk-head screws in a recess machined in the base of the router, so that the bush is always fitted concentric to the cutter. Other methods employ clamping screws (round-head screws that hold the edge of the bush) or bayonet fittings that simply twist and lock into the router base.

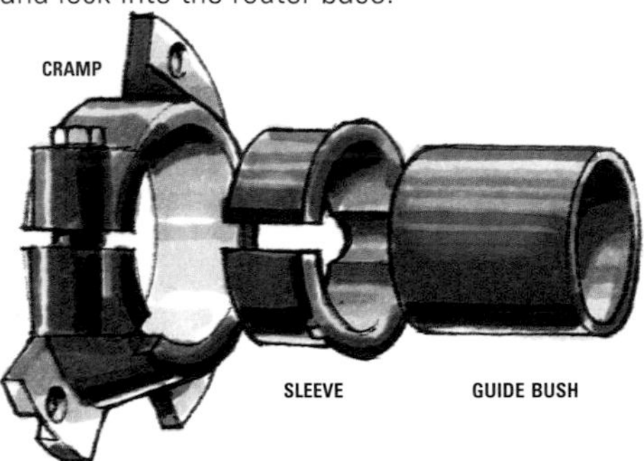

Adjustable guide bushes
Adjustable-depth bushes are fitted to some heavy-duty routers for applications that require a greater area of contact between the bush and guide edge or template; they prevent the spigot rising over the guide edge. The bushes are either held by a cramp fitted above the router base plate, or are threaded to screw into a holder fitted on the underside of the base plate.

GUIDE-BUSH MARGIN

Guide bushes are available in a range of outside diameters, not only to accommodate different sizes of cutter, but to allow the dimensions of the finished work to be varied by using different combinations of cutter and guide bush.

However, as the diameter of the cutter is always less than that of the guide bush, the difference between the outer edge of the bush and the tips of the cutter leaves a margin between the edge of the template and the actual line of the cut. This is referred to as the guide-bush margin. When using templates or positioning a guide edge, this margin must be taken into account.

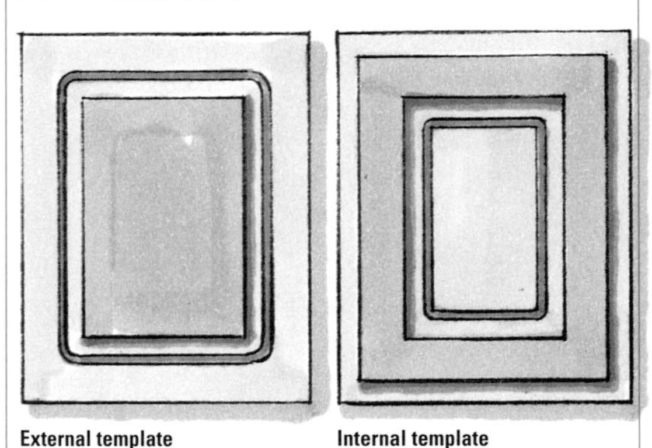

External template **Internal template**

Designing templates
An external template (where you cut around the outside) needs to be smaller by the guide-bush margin than the finished work. Conversely, an internal template (where you cut around the inside) needs to be larger by this amount. To calculate the guide-bush margin, subtract the cutter diameter from the outside diameter of the guide bush, then divide by two.

Cutter clearance

When selecting a guide bush and cutter, it is essential to allow adequate room for waste to clear freely between the cutter and the inside of the bush. This prevents waste from packing in the cutter flutes, causing the cutting edges to overheat. Approximately 3 to 4 mm (⅛in) clearance is recommended.

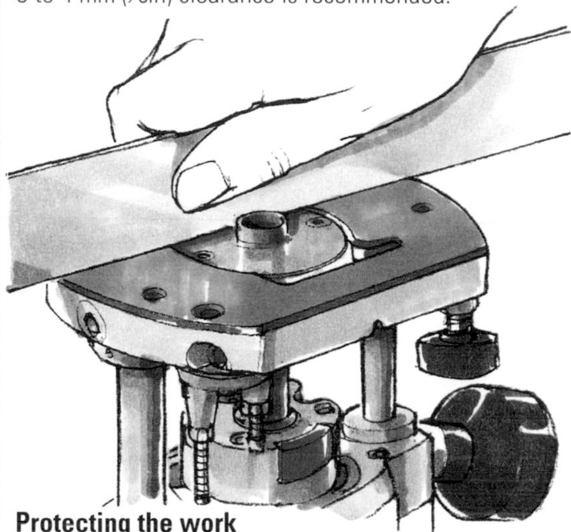

Protecting the work

Whichever type of guide bush is used on your router, always check that neither the guide-bush flange nor the screw heads stand proud of the underside of the base plate. In addition, check that the flange lies flat in its recess. Before fitting a guide bush, always clean wood dust and resin from the recess.

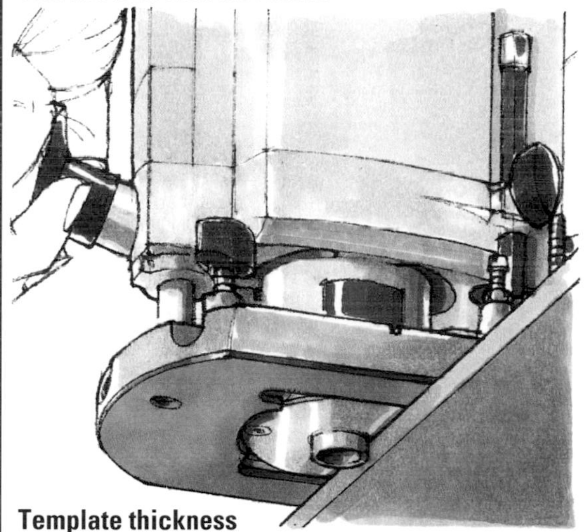

Template thickness

When using guide bushes, always check that the material used for a template or guide is thicker than the projection of the guide-bush spigot. If not, the tip of the bush is likely to mar the surface of the work and will tip the cutter out of true. For light or delicate applications, you can reduce the depth of the spigot by filing or grinding its bottom flat.

CUTTING INLAYS

Using various cutter/guide-bush combinations offers a simple, yet precise, method of cutting both the groundwork and the matching inlays for decorative and constructional applications. These include cutting out and patching knots and surface defects, as well as inlaying veneers and laminates.

For these applications, two guide bushes are used in conjunction with one cutter. A typical combination would be a 50mm (2in) guide bush, a 26mm (1in) guide bush and a 12mm (½in) cutter. For this method to work, the large-bush diameter minus the cutter diameter must equal the small-bush diameter plus the cutter diameter.

If the inlay material is valuable, use a small cutter, to minimize waste. You may need to finish the corners of either the inlay or the recess by hand.

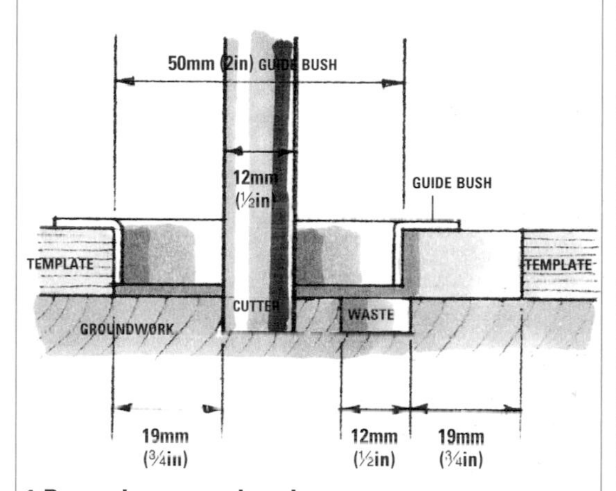

1 Removing groundwork

Use the larger guide bush for initial removal of the waste material or groundwork.

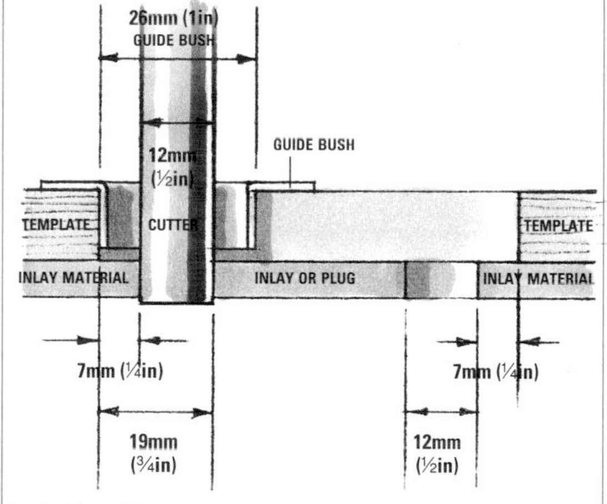

2 Cutting the plug

Fit the smaller guide bush to cut the matching inlay or plug from veneer or solid wood.

SELF-GUIDING CUTTERS

For many edge-trimming, moulding and rabbeting applications, self-guiding cutters offer the most direct method of guiding the router. Guided by a solid pin, ground on the bottom of the cutter, or by a ball-bearing race, fitted above or below the cutting edges, self-guiding cutters are generally used to follow the edge contour of the workpiece or a template mounted above or below it.

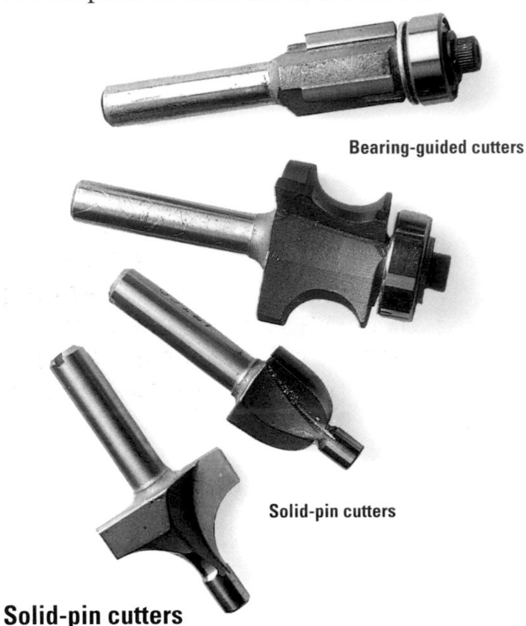

Solid-pin cutters
Solid-pin cutters are usually far cheaper than bearing-guided cutters, and the pilot pin can be ground to much smaller diameters than is possible with bearings. This allows them to cut close into corners. However, friction between the wood and the fixed pin is likely to cause them to overheat and scorch the surface of the wood. Pin-guided cutters are therefore more suitable for light applications and, in particular, fine or intricate work.

Bearing-guided cutters
Bottom-mounted guide bearings are usually fitted over a threaded pin ground on the end of the cutter and held in place by a small screw or nut.

When using bearing-guided cutters, always check that the bearing is secure before starting to cut, and take care not to put excessive sideways pressure on the bearing by forcing it against the template or guide edge. This can easily result in the bearing pin or screw bending and snapping.

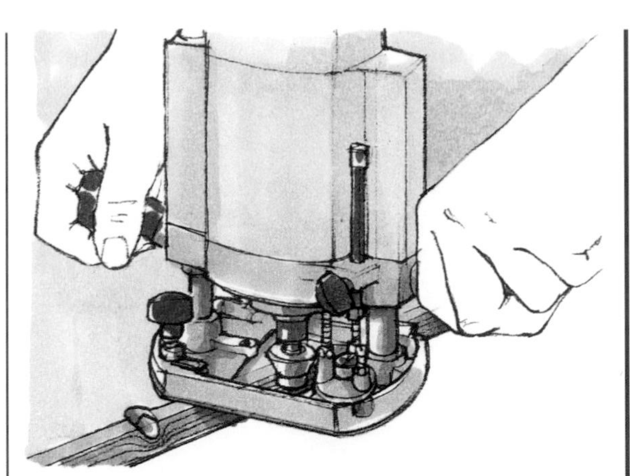

Edge finish
Guide bearings or pins will faithfully follow the guide-edge profile, reproducing any irregularities along the workpiece or template edge. Ensure that the workpiece or template edges are finished smooth and straight, and that high points, such as knots, spots of glue or dents, are smoothed off or filled, or they will be copied on the edge of the finished work.

GUIDE-BEARING MAINTENANCE

Avoid running guide bearings hot, as this melts the lubricating grease and causes the bearing to seize or break up. Don't attempt to clean a bearing by soaking it in a solvent, as this will dilute the grease, which is impossible to replenish successfully.

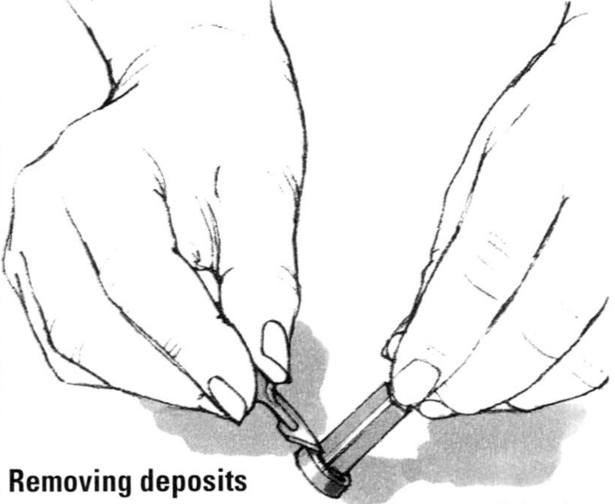

Removing deposits
Remove any deposits from the bearing by wiping it carefully with a rag dipped in solvent, or by gently scratching it with a sharp blade, taking care not to leave burrs on the surface of the metal. Always allow glues and adhesives used to lay laminates or veneers to dry thoroughly before trimming the edges, otherwise they can clog the bearing. Replace damaged or slack bearings by removing the retaining screw or nut, using an Allen key or spanner, then fit a new one.

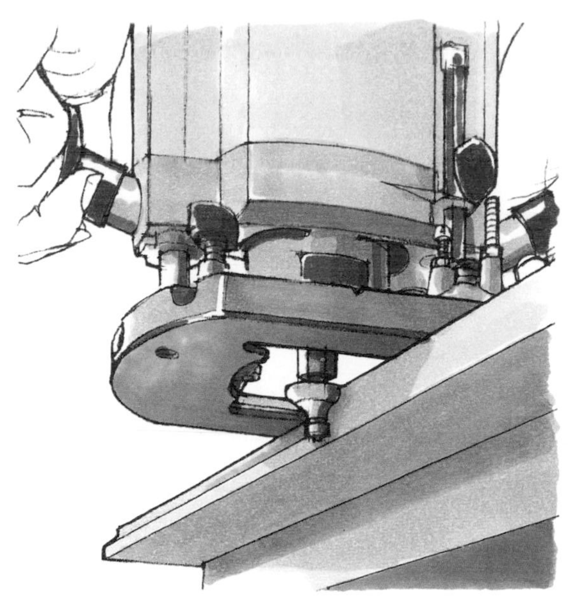

Using a plunge router

With plunge routers, use the same cutting procedures recommended for standard cutters, but ensure that the bearing has an adequate edge to run against. If the depth of cut reduces the guide-edge thickness to below 3mm (⅛in), clamp a deeper batten or template beneath the workpiece (see page 46).

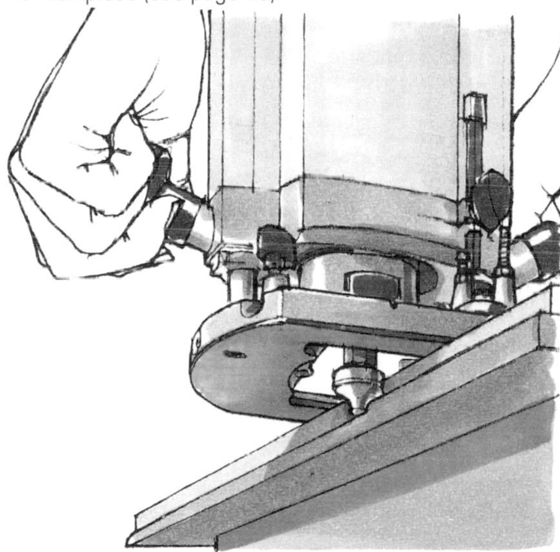

Shank-mounted guide bearings

When using shank-mounted guide bearings, set the depth of cut to allow the full bearing width to be used. When making shallow cuts, increase the thickness of the template or straightedge to provide an adequate running surface.

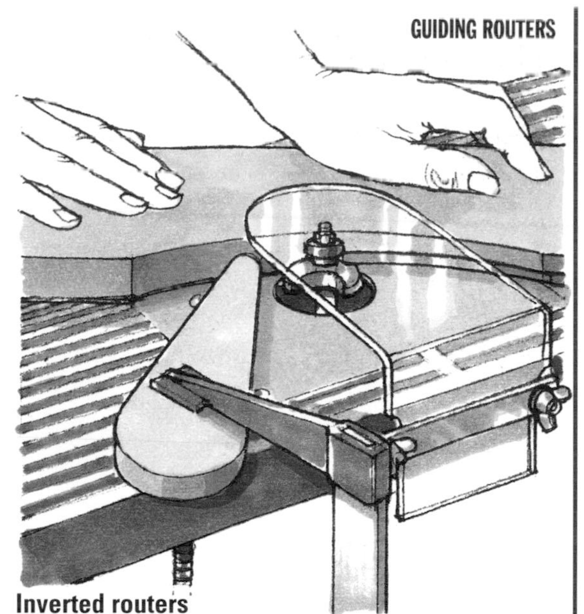

Inverted routers

Similar guidelines apply when using inverted or overhead-table routers, to avoid running a thin or feather edge against the bearing. For safety when using self-guiding cutters, retain the table fence and cutter guards. If this is not possible – when edge-moulding irregular-shaped work, for instance – fit a secure guard above the cutter.

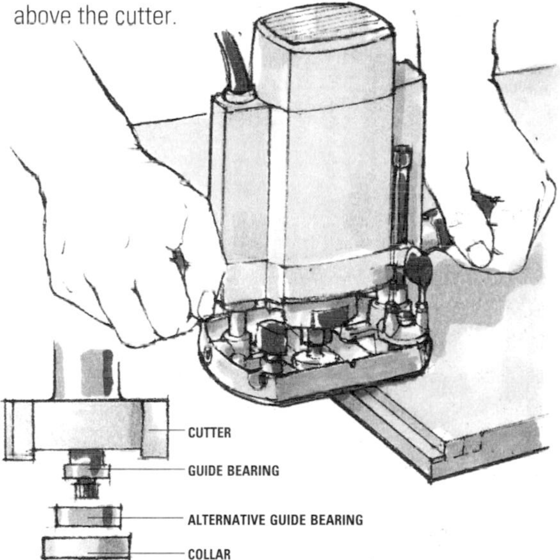

Rabbeting with bearing-guided cutters

Different-width rabbets, stepped rabbets or decorative stepped mouldings can be cut using the same cutter, simply by changing the bearing for one of a different diameter. Alternatively, metal or nylon collars can be fitted over a single guide bearing to increase its diameter. If you can't get collars from a stockist, you can turn them on a lathe; the inner diameter should be a press fit over a standard-size bearing.

CUTTING CIRCLES AND CURVES

Although you can use templates for cutting curved work, it is also possible to cut circles and arcs by swinging a router about a fixed point, using an adjustable trammel – a rigid rod or beam that incorporates an adjustable-height centre point. There are also trammels for cutting ellipses. (For cutting template-guided circles and irregular curves, see page 56.)

Since most router cutters can be used with trammels, you can cut plain or moulded circles and ellipses. Alternatively, cut a square-edge circle first, using a trammel, then mould it with a bearing-guided cutter.

Adjusting the trammel

Loosen the fence cramp to adjust the length of the trammel rod to match the radius of an arc or circle. When cutting external arcs, measure from the centre point to the nearest cutting edge. For an internal arc, include the entire cutter, measuring from the cutting edge furthest from the point.

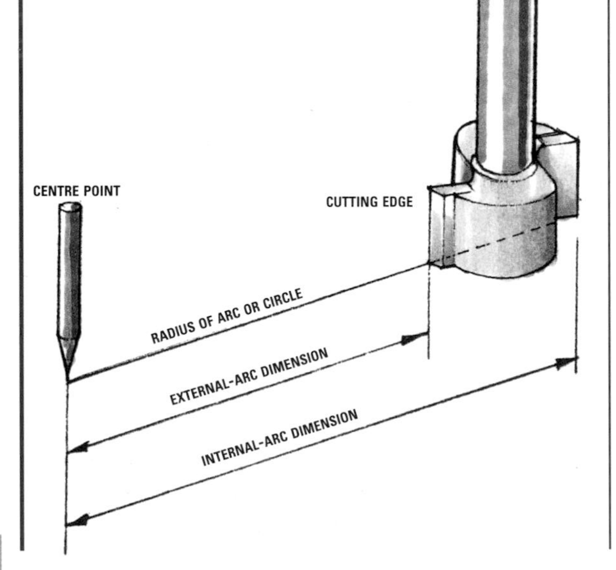

Single-rod trammel

The simplest trammel is a single rod held in one of the side-fence cramps on the base of the router. The centre point, usually mounted in a threaded hole at one end of the rod, can be raised or lowered as required.

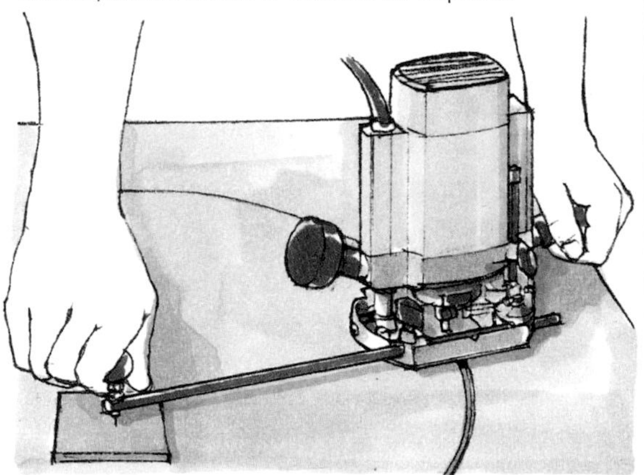

Locating the centre point

When cutting in waste material, or on a surface that can be filled or faced, the centre point can be pressed into the face of the work. When it is preferable not to mar the wood, press the point into a thin plywood or MDF pad stuck to the wood with quick-setting hot-melt glue or double-sided adhesive tape.

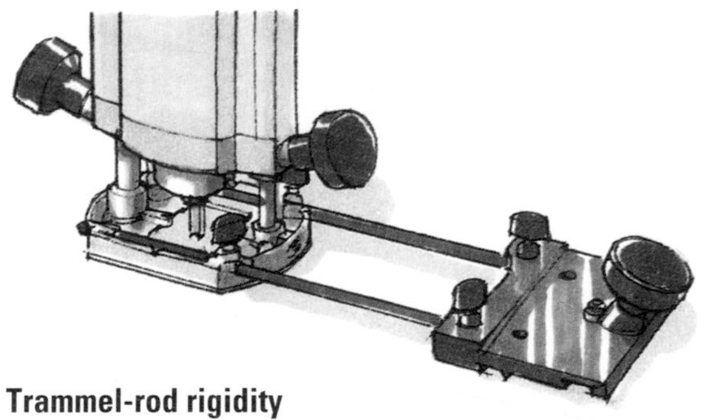

Trammel-rod rigidity
Ensure that a trammel rod is rigid enough to prevent flexing, since this causes variations in the radius of the arc as the router is swung about the centre point. When cutting large-diameter circles, use a twin-rod trammel or a flat beam that forms a sub-base to the router.

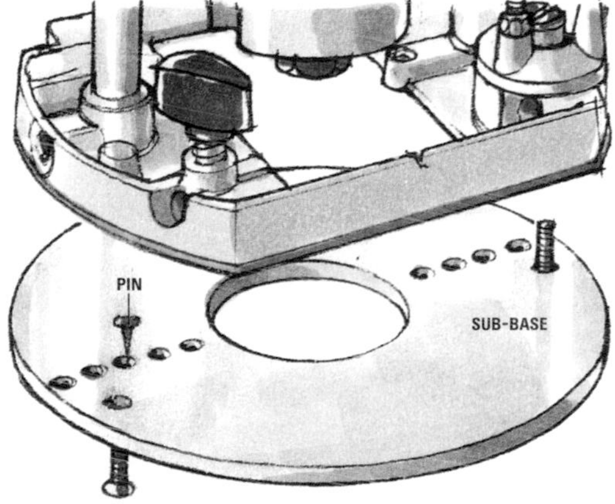

Cutting small circles
For cutting circles with a radius shorter than the width of the router base plate, attach a sub-base that incorporates its own centre point. Cut a disc from polycarbonate or shatterproof acrylic sheet, and drill a series of countersunk holes for a large-headed pin or tack, so that the arc radius can be adjusted in regular increments. With the pin in the required hole, screw the sub-base to the underside of the router.

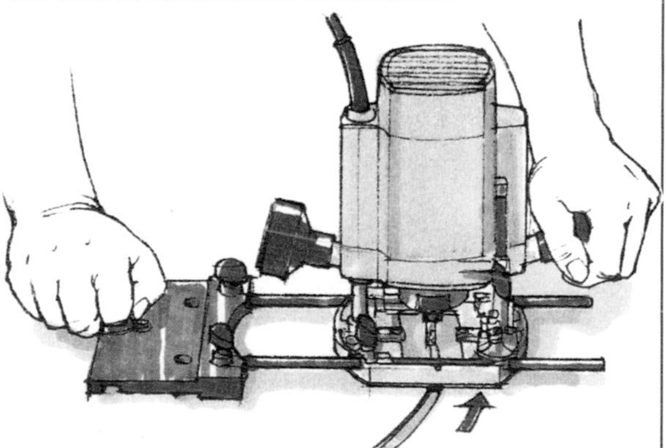

Working with trammels
When working with a trammel, keep the centre point pressed firmly into the work or pad. Feed the router in an anticlockwise direction.

CUTTING ELLIPSES

Ellipse trammels or jigs have two sliding pivot points, one to set the length of the major axis, the other to set the length of the minor axis.

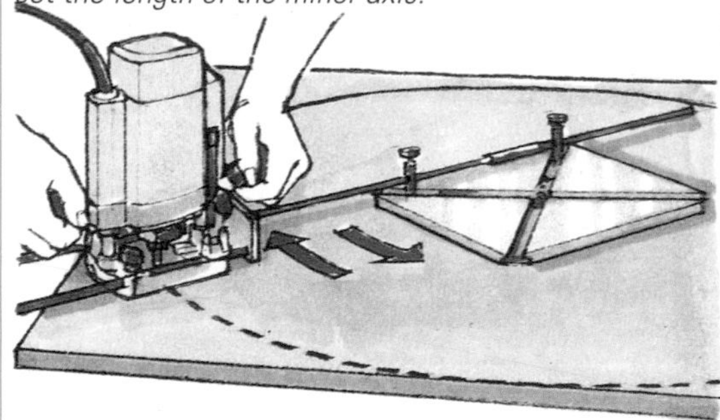

Direction of feed
Both pivots slide in T-slots or grooves cut at right angles to each other in a centre plate. The plate is fixed in the centre of the workpiece before the axis lengths are set out. The router is then swung about the plate to cut the ellipse. As both pivots are captive in their slides, the router can be swung in either direction when cutting an ellipse. Feeding with the cutter rotation (back cutting) can be beneficial, as it limits break out and leaves a smoother edge finish.

Making repetitive cuts
When making repetitive cuts from the same centre, the trammel point may enlarge the locating hole. You can avoid this in one of two ways.

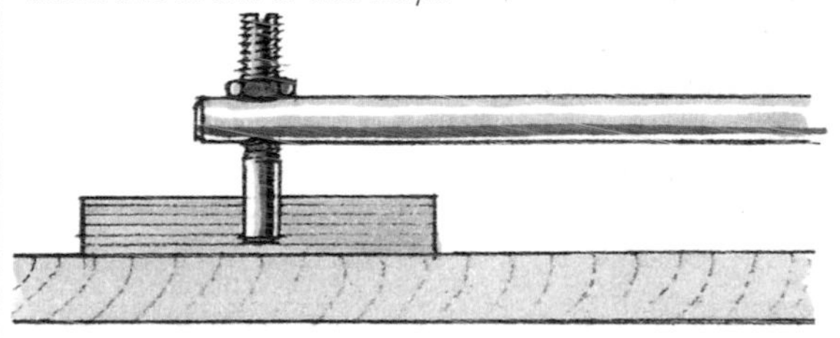

1 Using a threaded rod
Replace the threaded centre-point pin of the trammel with a partly threaded rod. Drill a similar-diameter hole in a pad and position this on the workpiece.

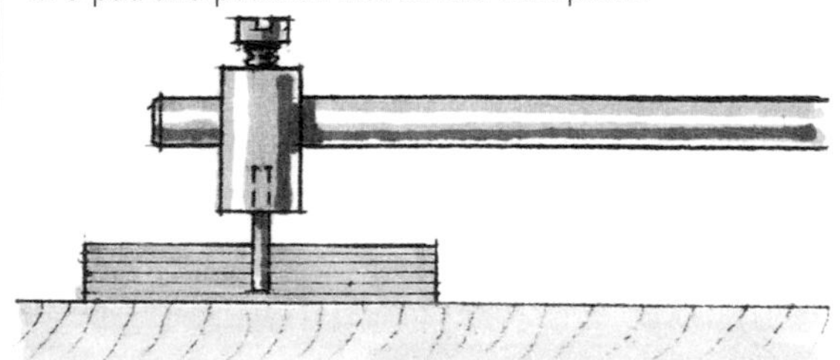

2 Using a block
Make up an adjustable block to be a slide fit on the trammel. Locate this over a loose pivot pin glued into a pad in the centre of the workpiece.

TEMPLATES

Using guide bushes or bearing-guided cutters, templates simplify demanding routing operations, such as decorative relief carving, and offer a quick and accurate method of performing repetitive tasks, such as cutting woodworking joints and hinge recesses. Working with templates saves time and reduces the risk of wasting expensive timber.

Templates are often incorporated into jigs, to be positioned quickly and accurately or to enable the simultaneous machining of several workpieces.

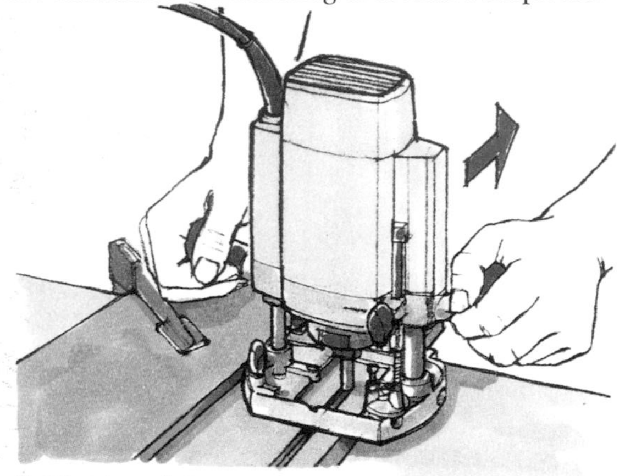

Cutting templates
Traditional woodworking handtools, electric jigsaws and other power tools can all be used for cutting templates. However, it is often worth using the router itself, as the template edges will be left clean and square, without needing further cleaning up.

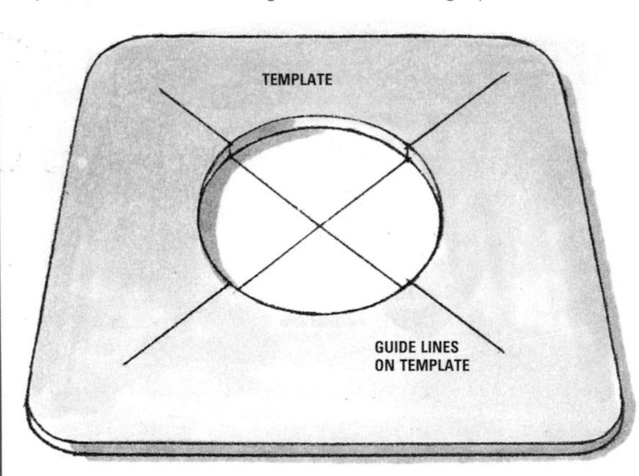

Registering a template
Locating battens, blocks or dowels, glued or screwed to a template, can be used to register it accurately on the workpiece. Alternatively, draw centre lines and registration marks on the template, to align with matching points on the work.

Template materials
Templates can be cut from any scrap piece of smooth sheet material that is easy to work and finish. This must be thick enough to keep a guide bush clear of the work, but not too thick for an adequate depth of cut. MDF is good for most applications, but for more-intricate or frequently used templates utilize a non-shattering plastic material such as polycarbonate.

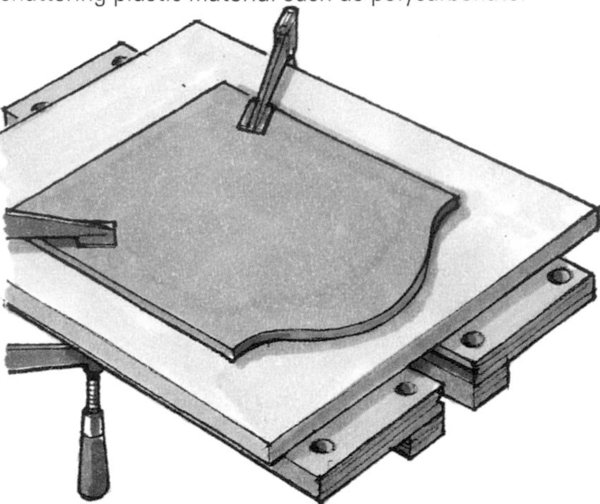

Securing templates
Always mount templates securely. Locating blocks and the downward pressure of the router may seem to hold a template in place, but it is always wise to secure it with cramps or other holding devices.

Templates can be pinned or screwed to rough, painted or concealed work. On flat work, they can be held with double-sided adhesive tape, as long as both surfaces are free from dust and grease. For stronger adhesion, seal the surface of the template (and the work, if possible) with a sanding sealer.

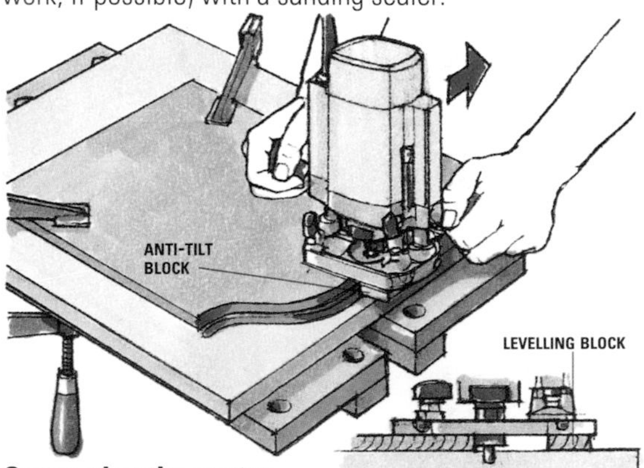

Supporting the router
Templates should afford as much support for the router as possible. Where the router overhangs the edge of a template, or where a template overhangs the work-piece, fit levelling blocks around the work or mount an anti-tilt foot or block on the router.

CHAPTER 6

One obvious attraction of power routing is the versatility of the tool. Equipped with a basic portable router, you can undertake a range of tasks, from ploughing simple grooves to boring holes. Yet this is just the beginning. With imagination and a little experience, you can cut ornamental mouldings, machine flutes or beads along tapered workpieces, and even carve or engrave decorative wall plaques.

GROOVES AND HOUSINGS

Grooves and housings are essentially trenches or slots cut into the face or edge of a piece of wood. A groove runs longitudinally, parallel with the grain, while a housing invariably runs across the grain. There is little to distinguish one from the other when machining a composite material like MDF, yet we tend to stick traditionally to the accepted terms, depending on how the trench relates to the proportion of a workpiece.

Grooves are used a great deal in woodworking to house cupboard backs or drawer bottoms. Door frames are also often grooved on the inside to receive fielded or plain panels. Housings tend to be used more for constructional purposes, to house fixed shelves or stair treads, for example.

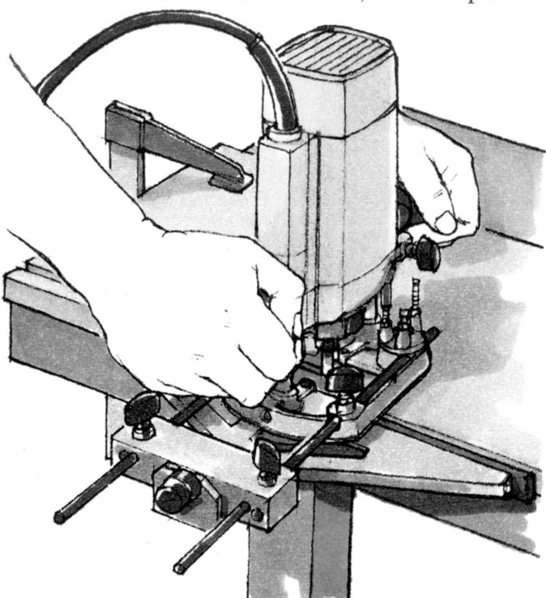

Cutting grooves
When machining grooves in the face of a workpiece, support the wood flat on the bench or across trestles. Ensure that cramps or the bench vice will not interfere with the passage of the router. While making the cut, hold the fence or router base securely against the guide edge, to prevent the cutter wandering off line.

To cut a through groove, plunge the router before feeding the cutter into the work. When making stopped cuts, either plunge the cutter on a pencil line that marks the extent of the groove, or fit wooden stops to a straightedge guide (see page 48).

WIDE GROOVES AND HOUSINGS

If the proposed groove, housing or slot is too wide to be cut in one pass, fit a smaller straight cutter and make the cut in two or more parallel passes. Make the first cut nearest to the guide edge, cutting to the full depth in stages. Adjust the side fence or straightedge guide to widen the trench, and repeat.

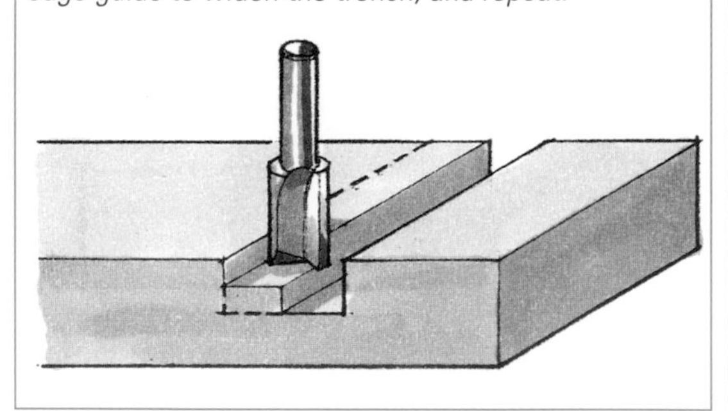

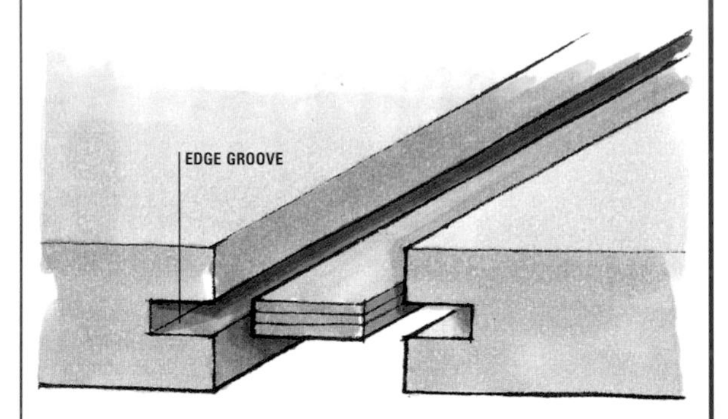

Cutting edge grooves
Boards can be joined edge-to-edge by gluing a plywood tongue into grooves cut along their edges. When using a standard straight cutter, clamp the workpiece on edge in the vice or, using G-cramps, against the edge of the bench. With long workpieces, arrange suitable support at the ends and along the length to prevent bowing. Twin side fences will help keep the cutter on line (see page 46).

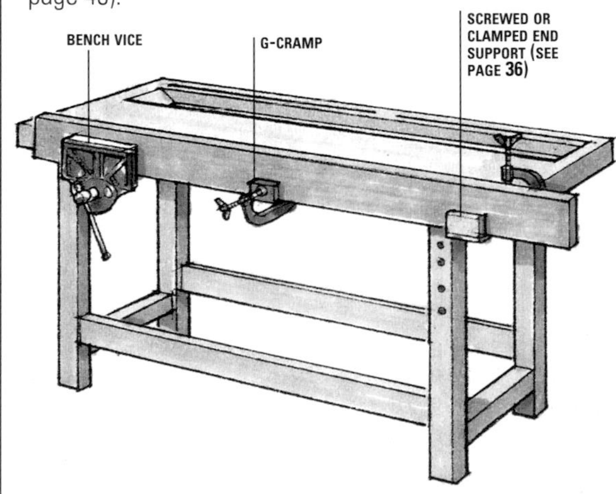

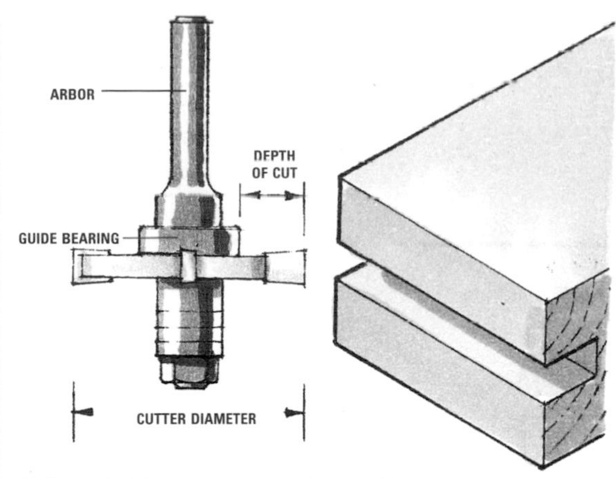

Using slotting-and-grooving cutters

Arbor-mounted slotting-and-grooving cutters, used singly or in combination, are designed for cutting slots or grooves along the edges of workpieces held flat on a bench, or a limited distance from the edge of a board held upright in the vice. The depth of cut is determined by the radius of the cutter minus that of the guide bearing, and is set by mounting the guide bearing above or below the cutters.

Although you can cut most slots and grooves in one pass using arbor-mounted cutters, you may need to use the side fence or a large-diameter guide bearing to limit the depth of the initial pass on hard or difficult timbers.

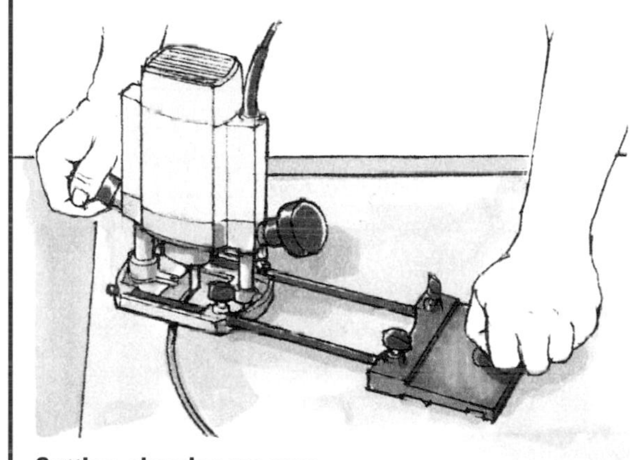

Cutting circular grooves

To machine a perfectly circular groove, for example to accommodate a desk tambour, attach a trammel to the router (see page 54). Secure the work on a flat surface and mark the centre of the circle. If the centre of the circle falls outside the area of the workpiece, temporarily glue or screw a block of wood to the bench, and use that to locate the trammel point.

Following irregular curves

To cut a groove that will follow an irregular curve, use a guide to track an external or internal template (see page 56). Always ensure that the template is held securely to the workpiece before cutting.

Cutting housings

When machining a housing across the face of a workpiece, use a straightedge or T-square to guide the router (see pages 47–9). To guarantee that shelf housings are identical on the two sides of a cabinet, either clamp both workpieces side by side and machine them simultaneously, or cut regularly spaced housings with the aid of a spacer batten (see page 47).

Cutting angled housings

Stair-tread housings, for example, are machined at an angle across the strings. Guide the router against a straightedge guide clamped at the required angle or, for greater precision and speed, make an adjustable T-square (see page 49).

PREVENTING BREAK OUT

Cuts made across the grain cause wood fibres to split out along the rear edge as the cutter breaks through. One way to avoid break out is to clamp a waste batten against the edge to support the fibres. Alternatively, if the edge of the workpiece is also to be rabbeted or moulded, cut the housing first and remove the fibres as you machine the edge in the direction of the grain.

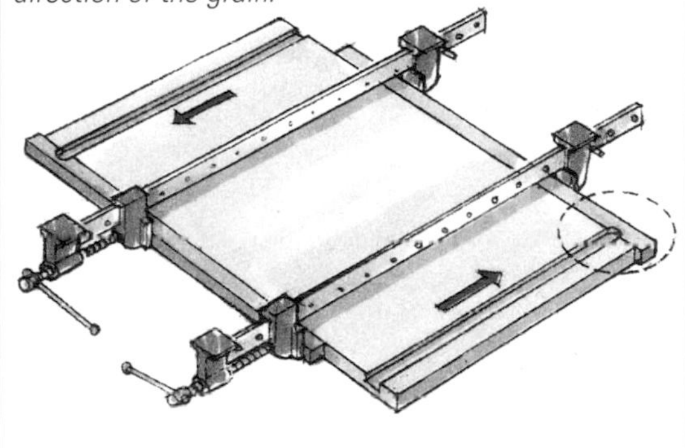

CUTTING RABBETS

A rabbet or step cut along the edge of a workpiece is another common device for joining a back panel to a cupboard or for fitting the bottom of a box. Cutting a rabbet also forms the basis for forming a tongue along the edge of a board (see page 103).

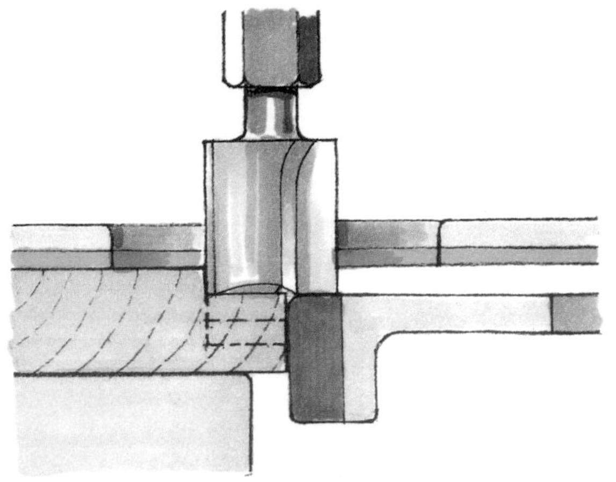

Using straight cutters
Fit a standard straight cutter with bottom-cutting edges. Ideally, this should be slightly larger in diameter than the width of the rabbet. Set the side fence to align the cutter with the inner edge of the rabbet, and cut in a series of steps until the full depth is reached. If you use a smaller-diameter cutter, first cut a narrow rabbet and then widen it with a second pass.

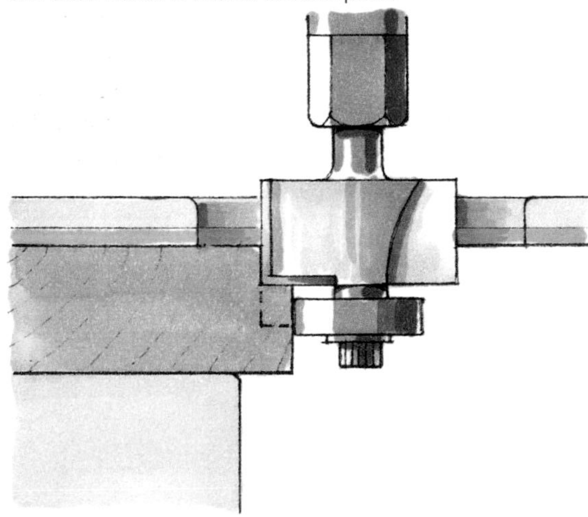

Using self-guiding cutters
Self-guiding rabbet cutters can be used on both straight and curved edges, but the width of the rabbet is restricted to available cutter and bearing sizes. Set the depth of cut for the first pass, and run the bearing against the edge of the work or a template fitted beneath it.

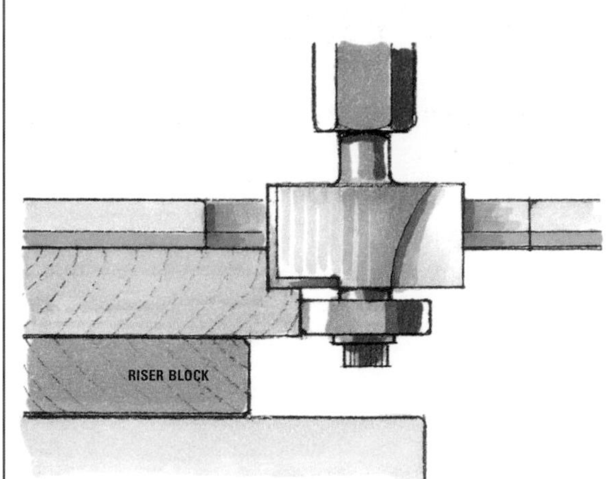

Providing bearing clearance
In most cases, clamp a workpiece that is to be rabbeted so that it overhangs the edge of the bench. If that proves to be inconvenient, such as when machining a curved or shaped edge, provide clearance for the cutter bearing by raising the work on a block or panel. Make sure the workpiece and riser block are firmly secured to the bench before cutting.

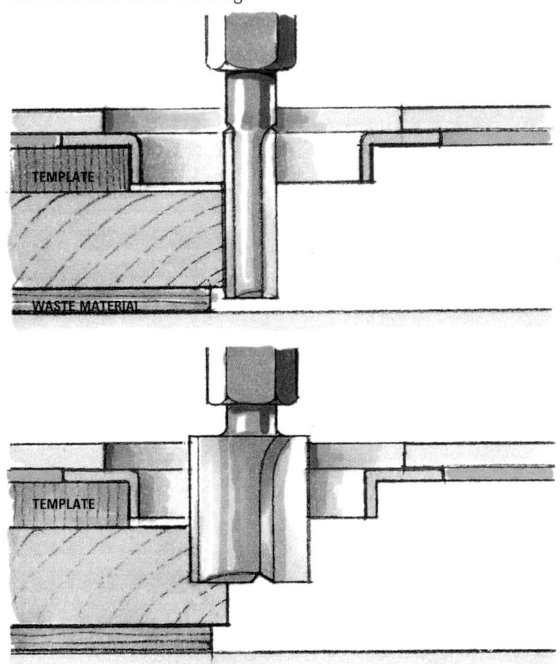

Using a template and guide bush
Select a guide bush and two straight cutters, one to trim the work, the other to cut the rabbet. Clamp the workpiece on top of a thin sheet of waste material. Fix the template to the workpiece, allowing for the guide-bush margin. Cut a perfectly square edge with the narrow cutter then, using the same guide bush, substitute the wider cutter to machine the rabbet.

MOULDING EDGES

Decorative mouldings are used extensively to finish the edges of table tops, panelled-door frames, chest lids and so on. Edge-machining operations are generally carried out using the side fence or a self-guiding cutter (see pages 17 and 44). When edge-moulding, it pays to fit a larger sub-base to the router, to minimize the risk of the tool tipping (see page 30).

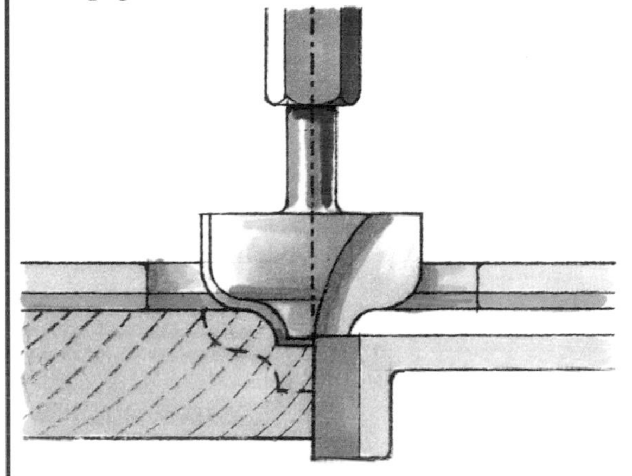

Moulding straight edges
Ensure that the edge of the work is finished square and smooth. Fit a plain moulding cutter and set the face of the side fence to align with the axis of the cutter. Adjust the depth of cut for the first pass, increasing the depth setting as the work progresses.

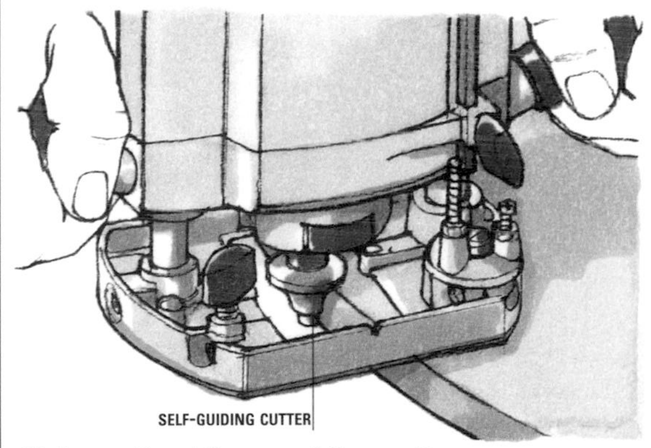

Using self-guiding moulding cutters
Use self-guiding decorative moulding cutters on both straight and curved edges. Change the diameter of the bearing to vary the profile (see page 25). Set the depth of cut for the first pass, and run the bearing against the edge of the work or a template fitted beneath it. Where the work cannot overhang the edge of the bench, follow the procedure described for self-guiding rabbet cutters (see opposite).

PRE-CUTTING EDGES

To reduce the amount of time that cutter edges are in contact with the material, use a jigsaw or bandsaw to roughly cut the workpiece oversize by up to 3mm (⅛in). Use a template cut exactly to the final shape to guide a flush-trimming cutter, to leave the edge square and ready for moulding.

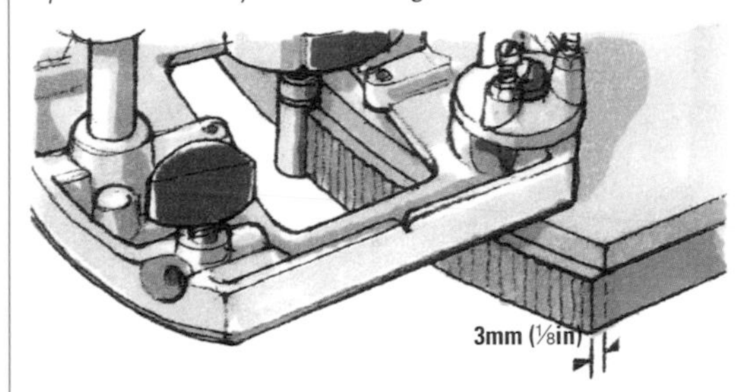

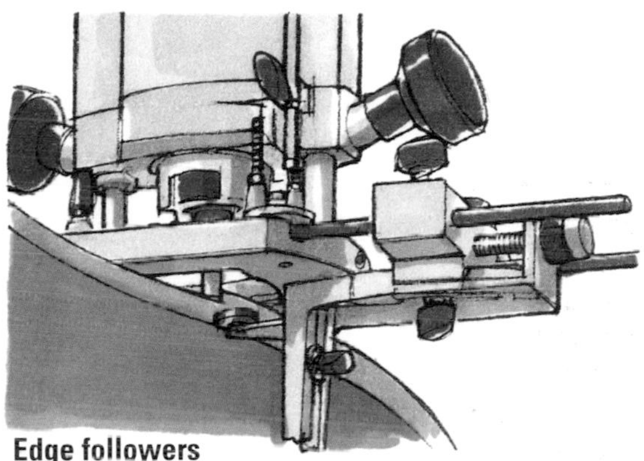

Edge followers
You can mould or rabbet any curvature, using plain cutters guided by an edge follower. This is a small-diameter wheel, attached to the side of the router by means of a bracket or the side-fence rods. The wheel is run along the edge of the work itself or against a template mounted beneath it.

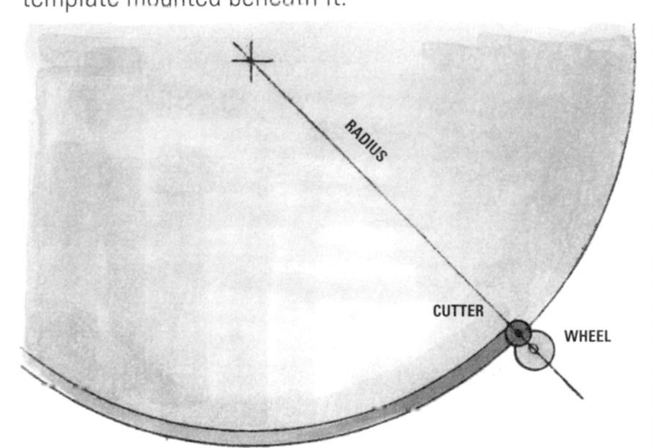

Aligning the edge follower and cutter
Throughout the operation, it is important that the axes of the cutter and wheel are kept aligned with the radius of the curved edge being worked.

BORING HOLES WITH A ROUTER

The majority of woodworkers would not consider using a router as an electric drill; yet, since the plunge carriage of the router is designed to enter and withdraw cutters at a precise 90-degree angle to the surface of the work, it bores perfectly accurate, clean-cut holes.

Don't use ordinary drill bits or masonry drills in high-speed routers. A range of special drills, dowel bits and plug cutters is available, as are straight and spiral cutters for drilling and boring. Short, slow-spiral dowelling drills are made for boring dowel holes. Whatever cutter you choose, always follow the manufacturer's recommended cutting speeds.

Hinge-sinking bit

Dowel drills

Drilling and boring cutters

All cutters used for boring must be ground with a bottom-cutting facility (see page 16). While you can use straight cutters, up-cutting spiral cutters clear the waste far more effectively. When using straight cutters, or large-diameter cutters that bore relatively shallow holes, plunge in stages, raising the cutter fully between each cut to clear the waste. It is also worth following this procedure with other cutters if the waste begins to pack in the flutes or if the cutter appears to be on the point of overheating.

CENTRING THE ROUTER

Although some cutters have a centre spur for accurate location on the work, you will need to centre square-ended cutters and drill bits by scribing centring marks on the router base plate, to align it with crossed lines drawn on the work.

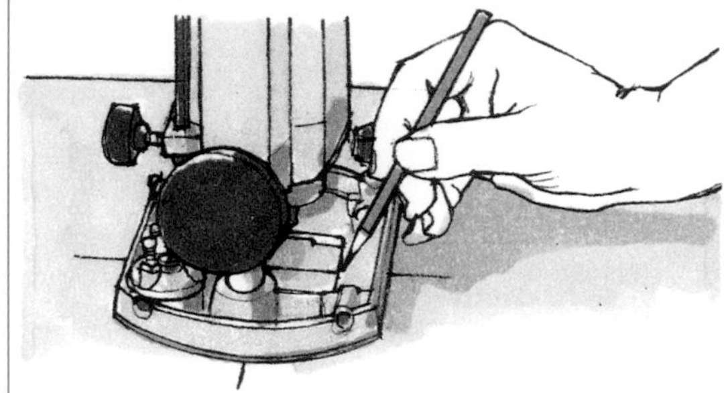

Marking the base plate

Unless your router already has centre marks cast in the base plate by the manufacturer, set out crossed centre lines at 90 degrees to one another on a flat surface. Fit a symmetrical pointed stylus in the collet and align the tip with the centre of the marked cross. Turn the router body to align the axis of the plunge columns along one centre line, then mark the four centring points on the router base plate. Check the accuracy of the marks before engraving them permanently, using a triangular file.

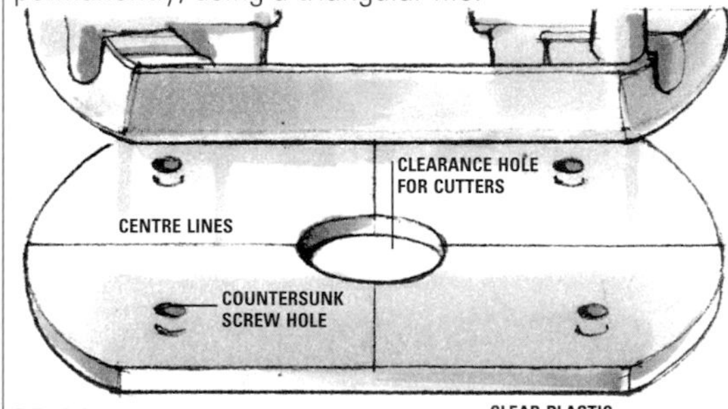

Making a sub-base

An alternative method is to fit a clear, shatterproof-plastic sub-base. Before cutting the material to size, scribe crossed centre lines at 90 degrees to one another. Fit a pointed stylus in the collet and apply double-sided adhesive tape to the underside of the base plate. Align the stylus tip with the centre point and press the router base down on to the plastic, then scribe the outline of the router base plate on it. Turn the router over with the plastic still stuck to it, and mark the centre of any mounting-screw holes. Remove the sub-base and cut it to size before drilling and countersinking the mounting holes. Check that the centre point aligns with the stylus point, then use a large-diameter straight cutter to gently plunge-cut through the plastic.

Locating the router

To prevent the router moving off-centre, stick a circle cut from medium-grade abrasive paper to the underside of the sub-base, using double-sided tape.

For more positive location, make a simple jig. Draw the outline of the base plate in the centre of a 300 x 300mm (12 x 12in) piece of 3mm (⅛in) MDF. Cut round the outline to provide a snug sliding fit for the base plate in the centre of the sheet. Place the sheet on the work, using crossed lines to centre it, then secure with cramps or edge battens.

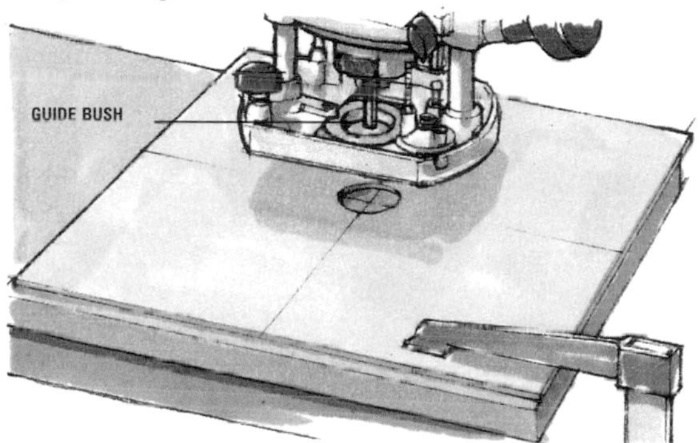

Using a guide bush

Alternatively, select a large-diameter guide bush and drill an equal-size hole through the centre of the jig. Align and secure the jig as before, then use the guide bush to locate the router on the work.

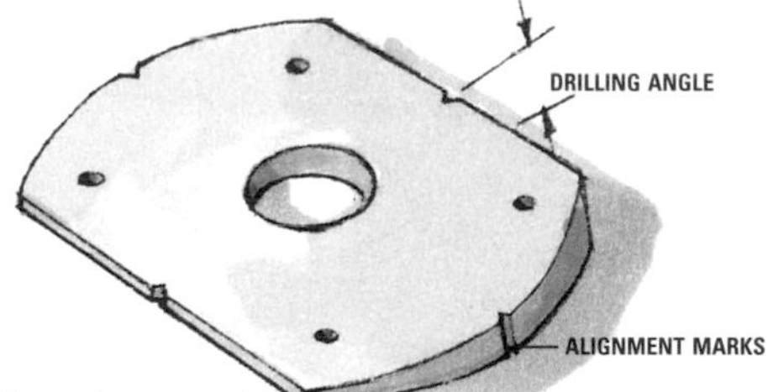

Drilling at an angle

To drill holes at a shallow angle, use a tapered sub-base. Cut the sub-base from thick board or plastic, then mark out and bore a large-diameter hole through the centre for the cutter. Cut to the required angle, and file aligning marks on each edge. To prevent the sub-base moving on the work, stick a sheet of abrasive paper to the underside.

COUNTERBORE BITS AND PLUG CUTTERS

Counterbore bits drill a clearance hole for a screw shank, a countersink for the screw head and a counterbored hole to sink the screw below the surface, all in one operation. Plug cutters are used to cut plugs from matching timber, for gluing into the hole to conceal the screw. The plug is then trimmed and finished flush with the surface. A similar counterbore bit without the countersink facility can be used for sinking square-head screws and bolts.
Plug cutters are also used to replace small surface blemishes with a patch of matching timber.

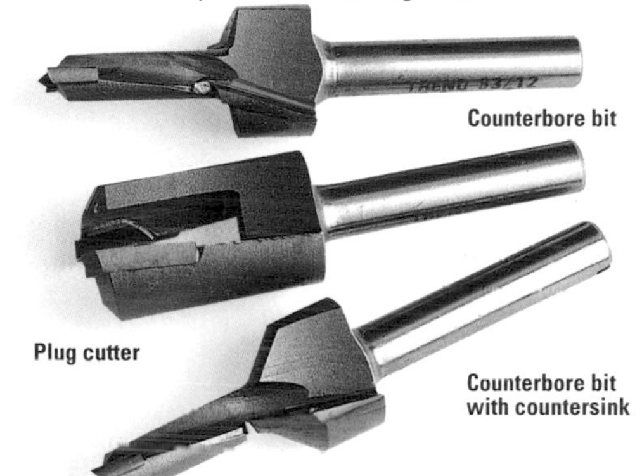

Positioning counterbore bits

The locating spur on a counterbore bit is sufficiently accurate for most purposes. However, for greater precision, make a locating jig (see left). When boring into curved or shaped surfaces, such as capping rails, fit a shaped mounting block beneath the jig.

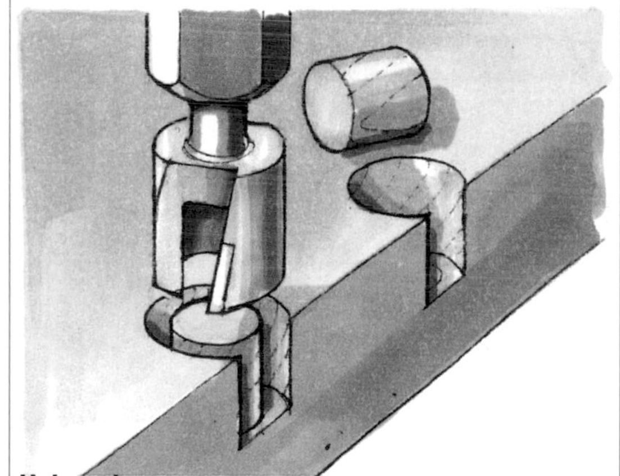

Using plug cutters

For safety, it is best to use plug cutters in a fixed-base overhead router, clamping the work securely to the table. However, if you have to use a hand-held plunge router, make a locating jig to prevent the tool moving as the cutter enters the wood. In either case, use only TCT plug cutters.

FLUTING AND BEADING

Cutting decorative fluting or beads along turned spindles, legs and columns is a relatively sophisticated routing operation, involving the use of a home-made box jig or a router lathe. It is also possible to turn a workpiece on the latter, but it may prove to be more convenient to use an ordinary woodturning lathe for the initial shaping, adapting a box jig to fit over the turned workpiece for fluting or beading *in situ*.

Making a box jig

To cut perfectly parallel flutes or beads along the length of table legs or spindles, make a simple box jig from 18mm (¾in) MDF or plywood. Join two 900 x 125mm (36 x 5in) side panels to 125 x 100mm (5 x 4in) end pieces, using corner rabbet joints. Brace the side panels with three MDF strips running from side to side across the base of the jig. Machine a 9mm deep x 5mm wide (⅜ x 3⁄16in) rabbet along the top inside edge of each side panel for the sliding router sub-base. Cut a 6mm (¼in) slot down the centre of each end piece, running 50mm (2in) from its top edge.

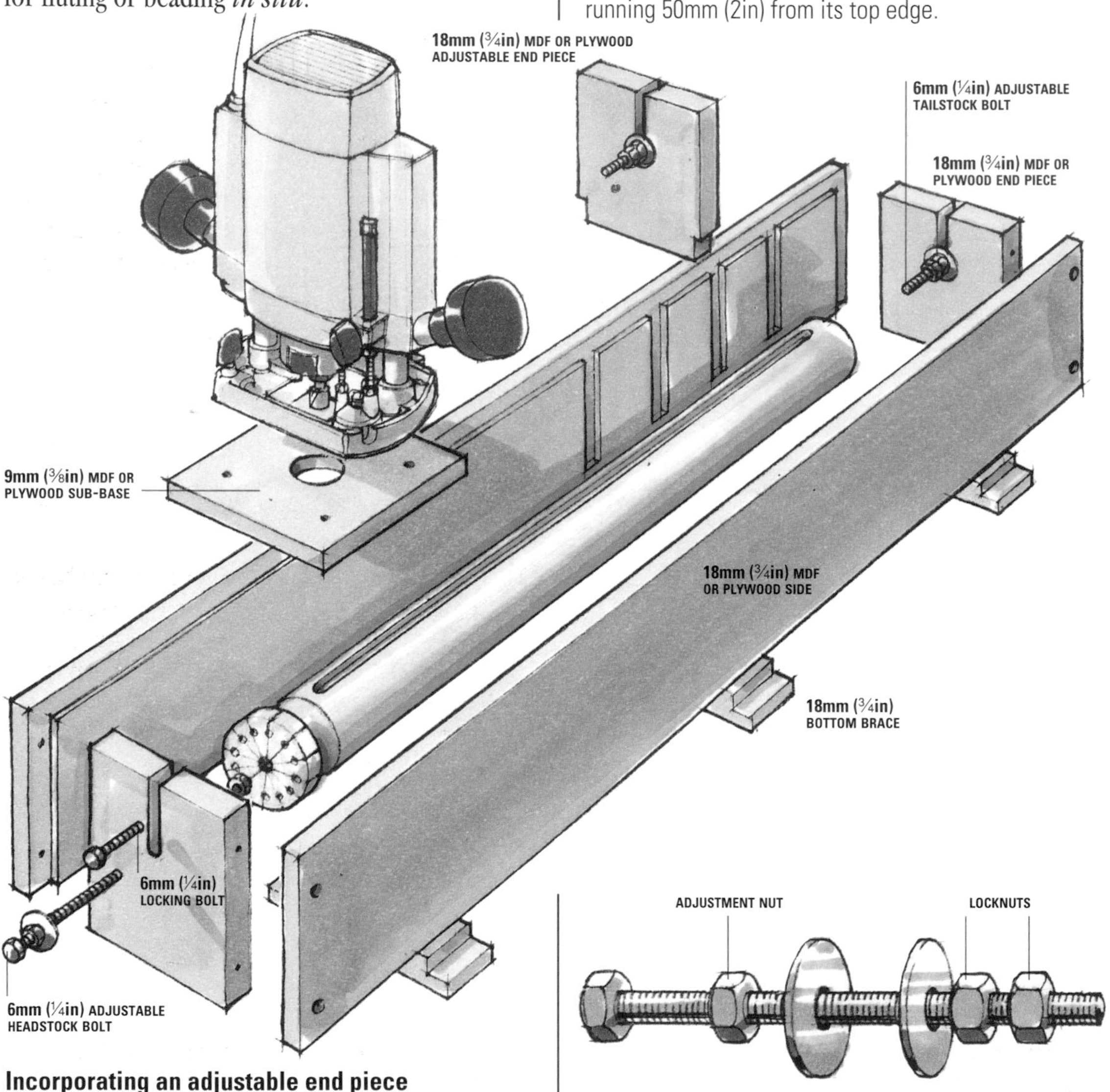

Incorporating an adjustable end piece

To make a jig that will accommodate a wider variety of workpieces, incorporate a movable, adjustable end piece that is located in vertical housings cut on the inside of both side panels.

Making the headstock and tailstock

Make an adjustable headstock and tailstock that will support the work at each end from 6mm (¼in) bolts. Each bolt needs two locknuts and washers.

Fitting the dividing disc
A 9mm (3⁄8in) MDF dividing disc is used to rotate the work through a number of identical increments. One of 12 holes spaced equally around the perimeter is used to lock the disc in the required position. Pinned to one end of the workpiece, the disc is mounted on the headstock bolt, and each time it is rotated, a separate bolt inserted through the slot in the end piece locates in one of the holes drilled in the disc. Mark and number the position of each hole on the edge of the disc for easy and quick reference.

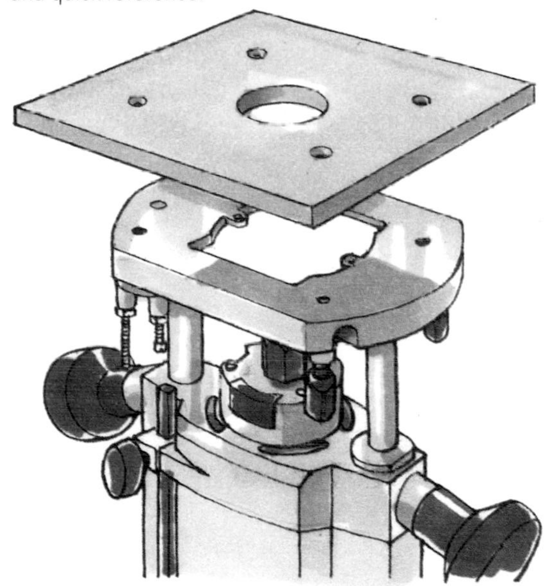

Mounting the router
The router is mounted on a 9mm (3⁄8in) thick sub-base, measuring 125 x 125mm (5 x 5in), that slides between the rabbets in the side panels. Bore a 40mm (15⁄8in) hole centrally for the cutter to pass through. Attach the sub-base to the router using the mounting holes in the base plate.

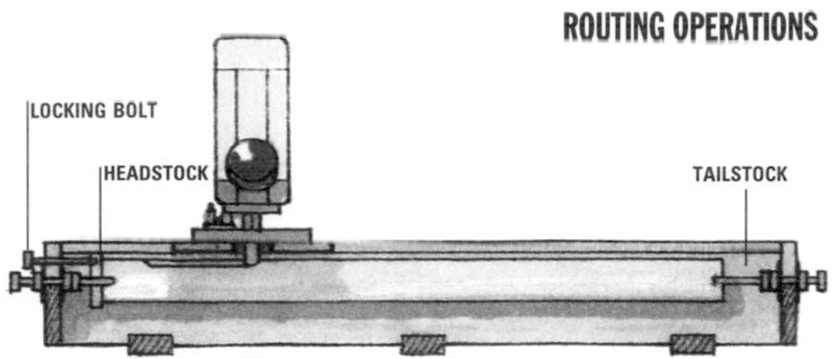

Fluting a turned workpiece
Turn the workpiece to size on a lathe, leaving an equal amount of waste wood at both ends. Drill a 6mm (1⁄4in) hole in each end for the tailstock and headstock bolts. Centre it with one of the bolts, then pin the dividing disc to the end of the workpiece.

Fit the work into the box jig and, once it is level, fix the headstock and tailstock in position with the locknuts. Insert the locking bolt to immobilize the disc.

Fit a cove bit in the router and adjust the cutting depth to machine the required fluting. Switch on, plunge the router and feed it to the far end of the work. Release the plunge lock and switch off. Release and turn the dividing disc to the next position, then repeat.

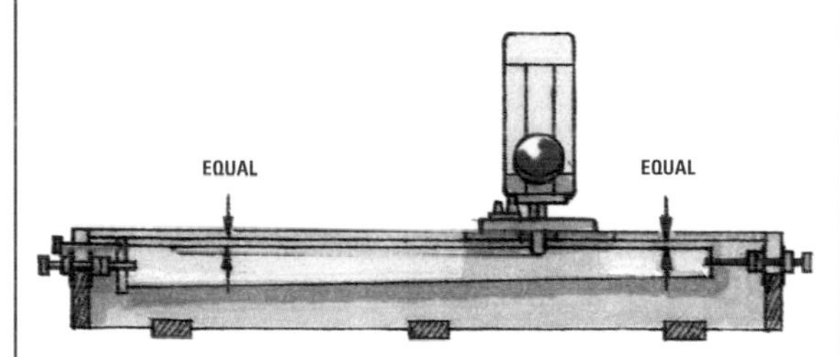

Fluting a tapered workpiece
In order to cut flutes longitudinally on a tapered chair or table leg, first draw the component full size so that all dimensions and angles can be measured accurately.

Set the work in the jig and adjust the height of the tailstock and headstock until the surface of the work is parallel with a straightedge laid across the top of the box. Tighten the locknuts.

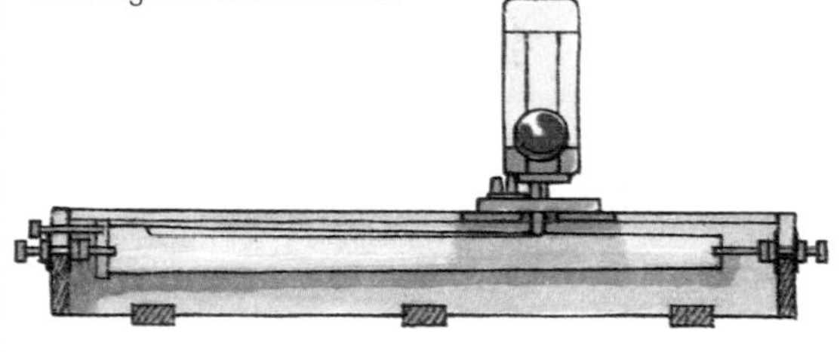

Tapered fluting
Fluting on a tapered leg looks best when the flutes themselves also taper, becoming narrower towards the smaller end of the workpiece. To achieve this, measure the depth of cut at the wider end of the leg, slacken the locknuts at the other end and lower the tailstock by about half the cutting depth. Tighten the tailstock nuts, and make a trial cut to check the results.

ROUTER LATHES

A router lathe is a factory-made accessory that will produce the same results as a box jig, but with much greater precision. It can also do many more operations, including barley-twist, diamond and hatched fluting. In addition, you can use the lathe to turn workpieces with the router mounted on a sliding plate.

Straight and spiral flutes and beads

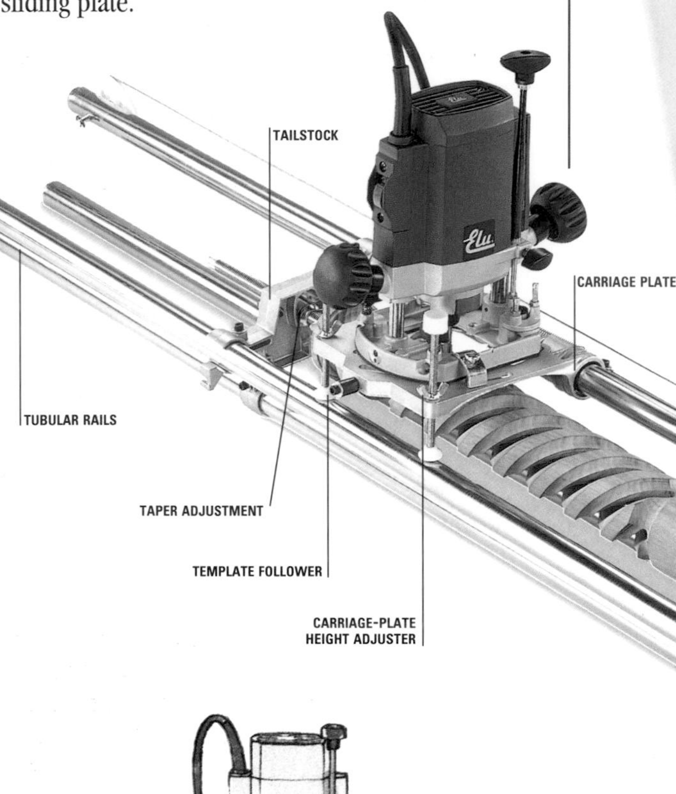

TEMPLATE

TEMPLATE FOLLOWER

Copying work

You can use a router lathe to reproduce any number of identical pieces from a template mounted on the front edge of the machine. A follower on the router plate runs against the shaped edge of the template as the router is fed along the lathe.

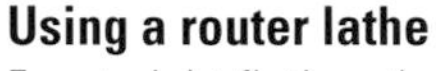

Using a router lathe

For straight fluting, the work remains stationary and the router is wound from one end to the other, using a cranked handle on the headstock. Adjustable stops limit the extent of travel.

When machining barley-twist fluting, the work rotates at a precisely governed speed that corresponds with a given router feed rate. This ratio is set to give a pitch angle for barley-twists of 45 degrees, and both right- and left-hand pitches can be cut.

FREEHAND ROUTING

Once you have machined the edges of a lock or hinge recess, or perhaps cut the shoulders of a tenon or lap joint, it is a simple task to remove the remaining waste wood, using the router freehand. However, far from being restricted to mundane tasks, a freehand router is a creative tool, much used by woodcarvers for roughing out prior to finishing with hand work, and as an engraving tool for lettering and other low-relief carving.

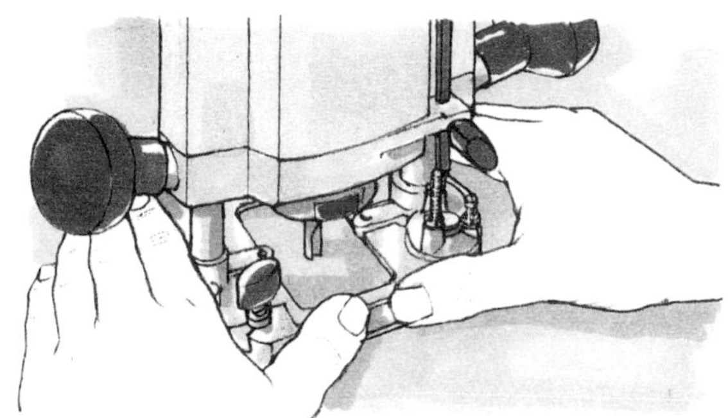

Controlling the router
The direction of rotation of the cutter tends to pull it sideways into the wood. This can be used to advantage when the side pull from the router is restrained by a guide or fence, keeping the router on course. When routing freehand, however, you have to anticipate this tendency for the cutter to wander and, since the cutter is also deflected by tough patches of grain, knots and fissures, you need to brace the machine against any sudden changes in direction. Hold the router by its base plate, keeping your hands lightly in contact with the surface of the work.

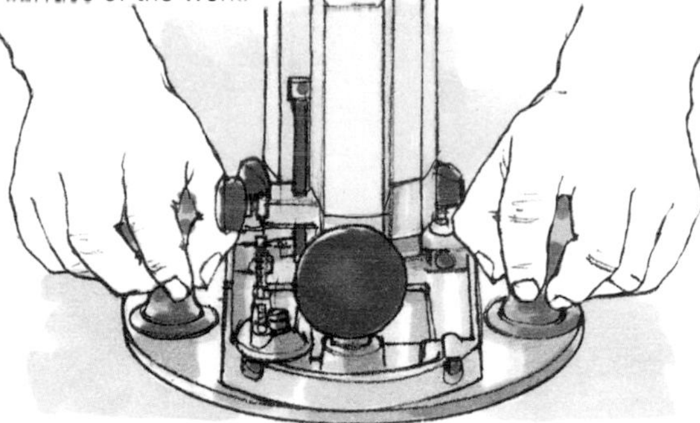

Fitting a sub-base
You can increase your control over the router with a clear-plastic or MDF sub-base fitted with two side handles (see page 30). Cut a 40mm (1⅝in) hole through the centre, and drill and countersink holes to take the mounting screws.

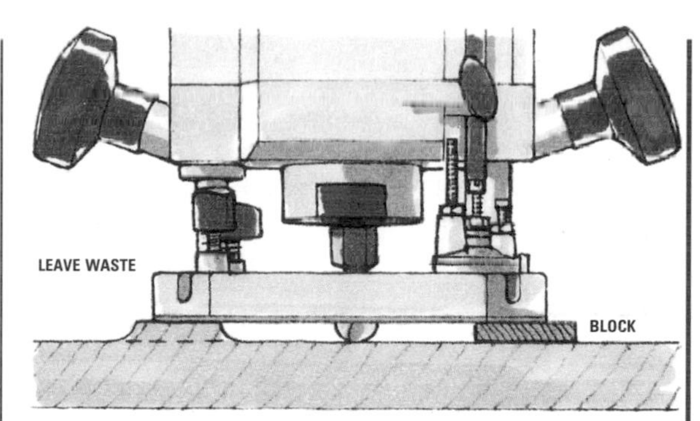

Supporting the router
To support the base plate, work out in advance a cutting sequence that will leave some areas intact until most of the waste has been removed. Where this is not practicable, fit an anti-tilt foot to the router base or stick blocks temporarily to the work with double-sided tape (see page 56). Alternatively, make an asymmetric sub-base that will help balance the router when machining close to an edge.

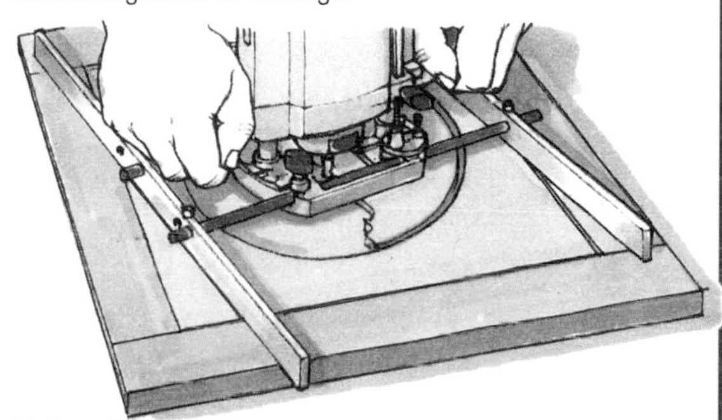

Using ski bars
Ski bars are designed to fit onto the side fence support rods, supporting the router over a large area when relief carving or removing waste from a wide pocket. The bars bridge between battens fixed all round the workpiece, or they can be fitted with nylon shoes that run on the bench top.

CUTTERS AND BURRS

A range of cutters is available for freehand work. Square-end cutters leave a flat surface, while V-groove and cove cutters can be used to texture it. In addition, rotary burrs and rasps are useful for shaping and smoothing wood. Try to follow the normal feed direction – against the rotation of the cutter – and cut in very shallow steps.

Rotary burrs

CARVING AND LETTERING

The router lends itself to freehand or template-guided lettering; two popular ways of creating decorative signs, such as nameboards for houses and boats. Any standard router can also be used to rough out low-relief carving, but a lightweight, detachable-base router becomes a hand-held power tool for shaping sculpture in the round. Use only burrs or rasps for this type of work, since other cutters are dangerous.

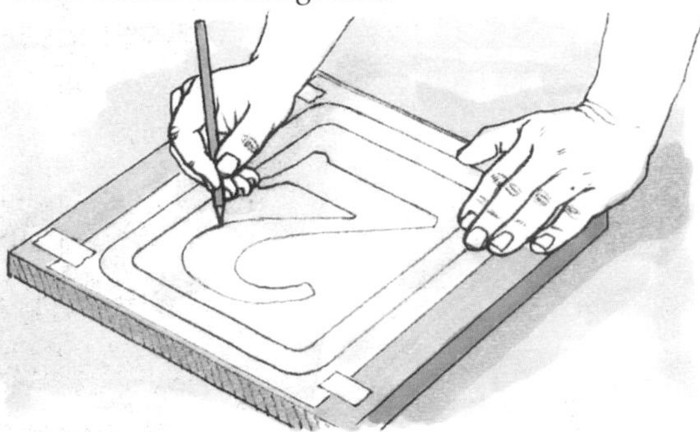

Preparing artwork for lettering
Lettering can be cut in relief, by removing the background, or the letter forms themselves can be recessed and borders or decorative motifs added. Draw your proposed design on paper, taking care to space out individual letters correctly. Alternatively, you can buy sheets of self-adhesive lettering, available in a wide range of typefaces. If you lack the confidence to prepare your own artwork, hire a graphic designer to create a pattern from which to work.

Trace off the finished layout. Transfer it to the work by first rubbing the back with soft pencil then, with the pattern taped in place, trace off the lines to offset them on the surface. On dark woods, you can use a photostat pasted to the surface.

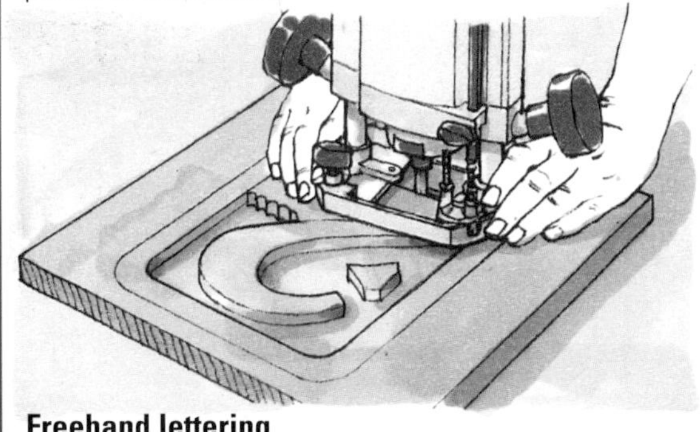

Freehand lettering
Follow the outlines of letters and other motifs marked on the work, using the procedures outlined for controlling the router (see page 67).

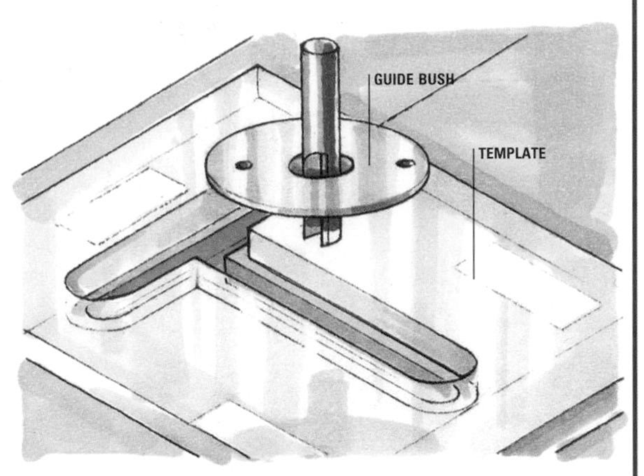

Lettering templates
Some woodworkers prefer to cut intricate shapes such as letter forms, using templates to guide the router. Cut your own from 6mm (¼in) thick plastic, or use sets of ready-made templates available in a number of suitable typefaces. Fix them in place with double-sided tape.

CARVING WITH A ROUTER

Trace out the design or draw it onto the surface of the work, hatching with a pencil to identify areas that might be of a similar depth or texture. Fit a coarse burr or rasp in the router collet and, holding the router motor housing in both hands, cut away large recessed areas first. Change to a finer burr for finishing. Clear waste material regularly from between the teeth of the burr, as this affects the surface finish.

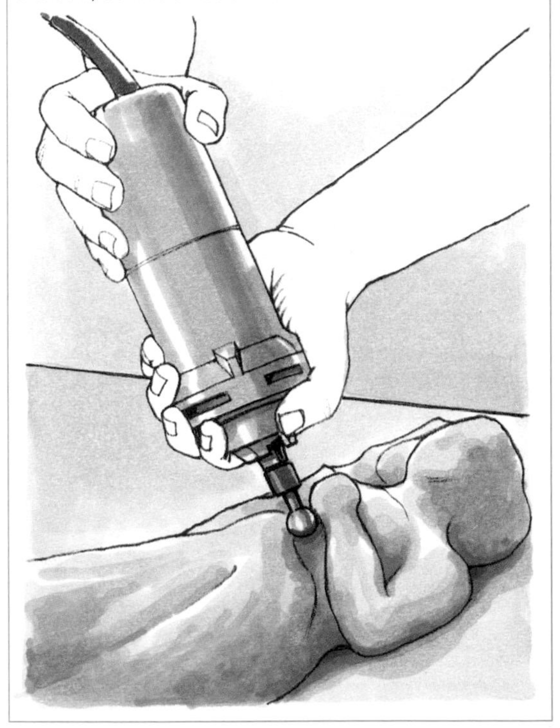

TABLE ROUTING

Chapter 7 Table routing offers a higher degree of precision, control and safety than you can achieve using a hand-held router. There are special-purpose routing tables and workcentres, or you can adapt a radial-arm saw for routing. With a table-mounted router, you can use relatively large, complex cutters; however, a variable-speed router is essential in order to achieve the necessary low cutter speeds.

INVERTED-ROUTER TABLES

A basic inverted-router table consists of a work-support table, router-mounting plate and a straight table fence. Other essentials are adequate guarding and, for ease of working as well as safety, hold-down cramps.

The support for the table can be a floor-standing cabinet or simply a low stand for clamping onto a folding portable workbench. Whichever method you employ, ensure there is easy access for adjusting the router.

ATTACHING FACES TO THE FENCE

Adjustable auxiliary faces attached to the fence prevent the work from dipping into the cutter at each end of the pass (see page 44).

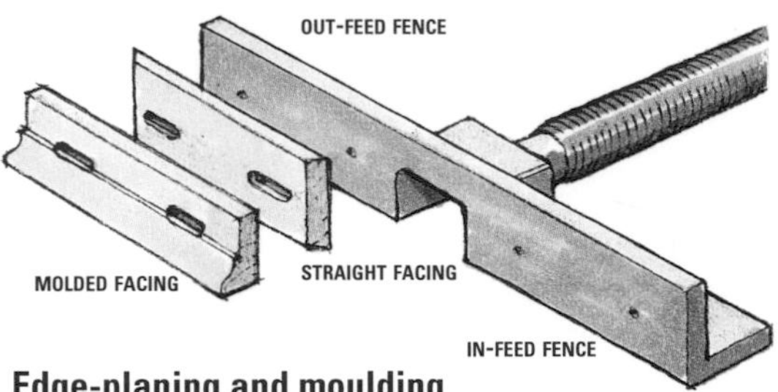

Edge-planing and moulding

Fitting a shallow auxiliary face to the out-feed side of the fence supports the edge of the work when edge-planing. Similarly, mount a shaped face when cutting full-width edge mouldings (see page 46).

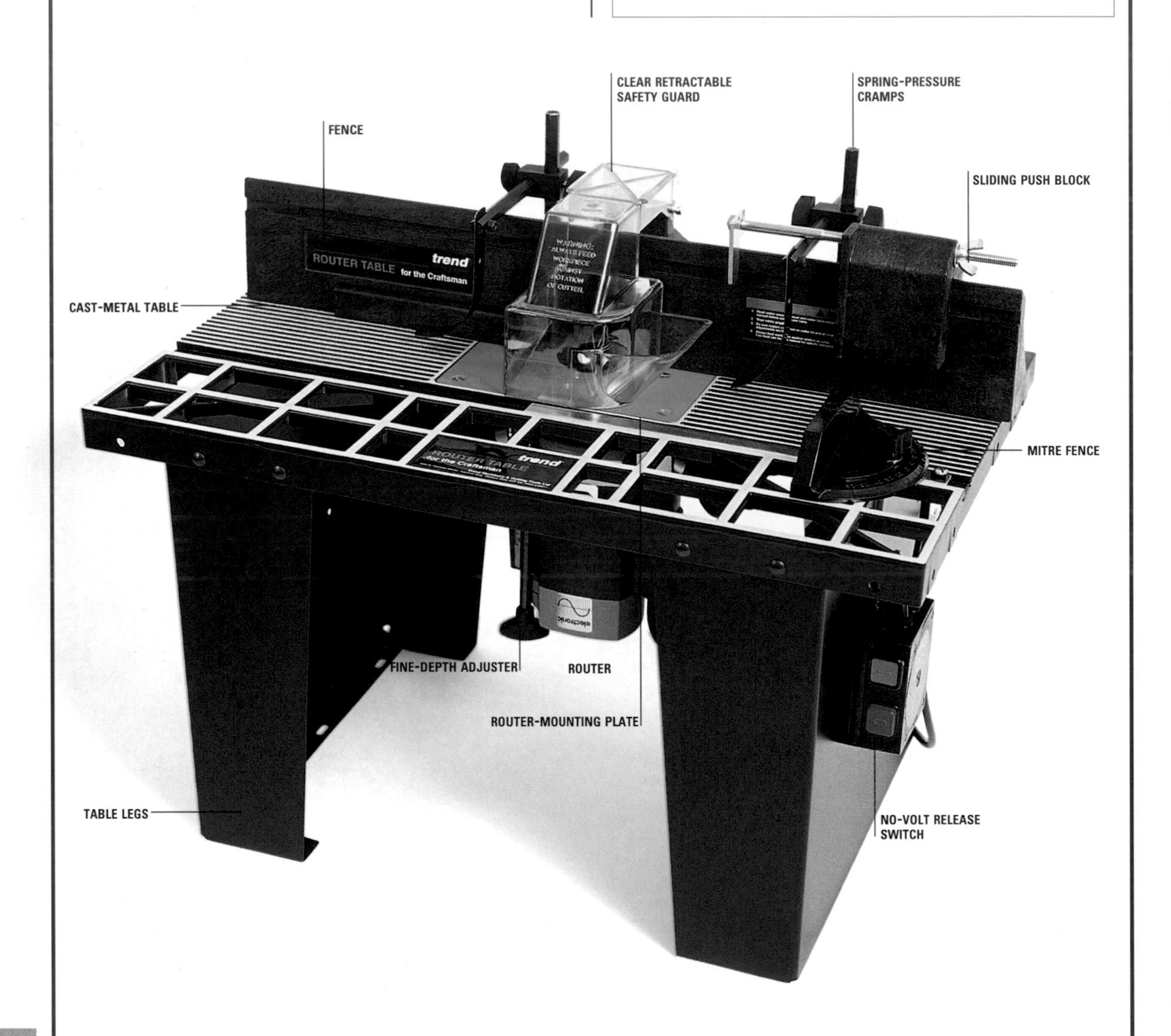

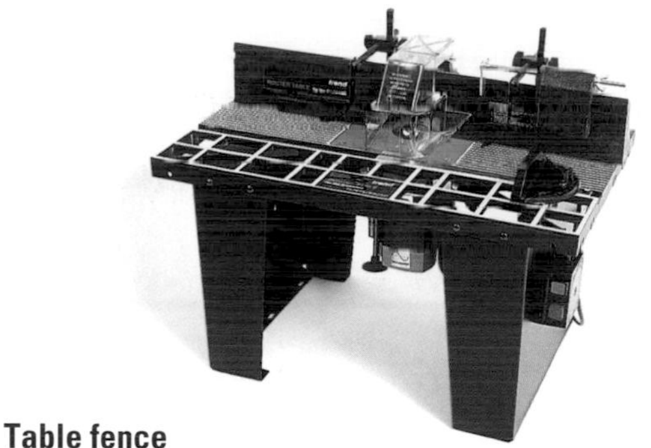

Table fence
The table is fitted with a rigid guide fence. The fence should have sufficient fore-and-aft adjustment to enable you to set the centre line of the cutter behind the face of the fence for moulding the edges of workpieces, and also to set the cutter a reasonable distance from the fence for cutting grooves and housings. Set the fence parallel to the table edge, to align it accurately with sliding mitre fences, work-holders and other jigs and guides.

Dust extractor
It pays to fit a dust-extraction port to the table fence. The port should be positioned directly behind the cutter recess and be hooded for maximum efficiency.

Spring-pressure cramps
These devices prevent the workpiece lifting from the table surface.

Mitre fence
A sliding mitre fence permits you to machine the ends of narrow workpieces.

Sliding push block
A sliding push block mounted on the fence allows you to perform a similar task with the workpiece held in a vertical position.

Remote switch
Avoid having to reach under the table, to turn the router on and off, by mounting a remote switch to one side of the table stand. No-volt release switches are best for this purpose – should the power supply be cut off, the router will not operate until you press the start button. It should also incorporate an accessible knock-off switch to cut the power in an emergency. No-volt switches are available from router-accessory shops.

Fine-depth adjuster
Since the plunge action is not required for most operations when using the router inverted, fit a fine-depth adjuster to allow precise setting of the cutter height above the table.

CUTTER GUARDS

As with all power and machine tools, adequate guards must be fitted to the router to prevent accidents. The dust-extraction hood on most table fences forms a guard to the back of the cutter. However, the exposed cutter in front of the fence face must also be guarded when routing. No part of the hand should be allowed to get closer than 25mm (1in) to the rotating cutter. ***In some illustrations, guards and hoods have been omitted for clarity.***

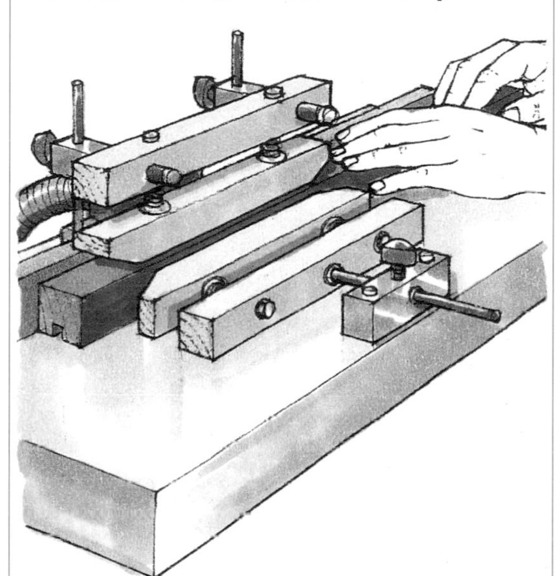

Hold-down guards
Hold-down guards not only cover the cutter but also act as spring-pressure cramps. Side-pressure guards protect the cutter from the side and hold the work against the fence.

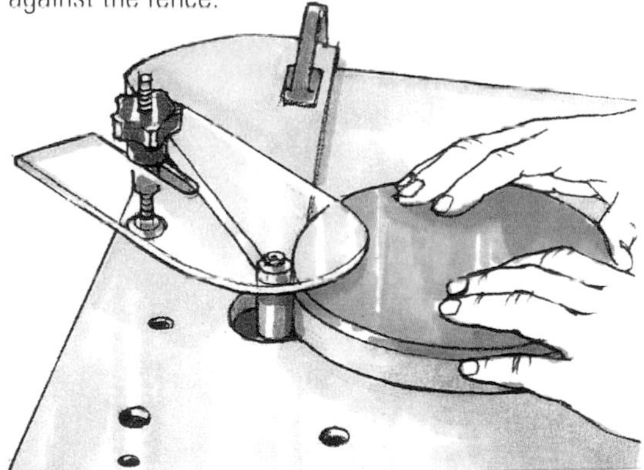

Top guards
Where hold-down guards are inappropriate, mount top guards over the cutter. These can be secured to the table fence or to a rigid bracket. If you use transparent guards, ensure they are made from a shatterproof plastic, such as polycarbonate.

MAKING A WORKTABLE

Although most proprietary router tables are made of aluminium or steel, man-made boards, such as MDF and dense particleboard, are perfectly adequate for a home-made routing table. Plastic-faced board is particularly suitable, as it is stable, easy to work, and has a relatively low-friction surface. It is also inexpensive – offcuts of kitchen worktop are ideal. Make sure there is a balancing laminate on the reverse side to prevent bowing. Mount the table on a sturdy underframe, or make a box from man-made board.

Mounting the router

Most router bases are drilled and threaded to take fixing screws for table and jig mounting. One of the advantages of using a thin metal table is that you can retain maximum cutting depth. If you use thicker table materials, you will have to set the router into the thickness of the board, rather than simply bolting it on. An alternative method is to cut a hole larger than the router through the table, and mount the router on a metal plate inset flush with the table surface.

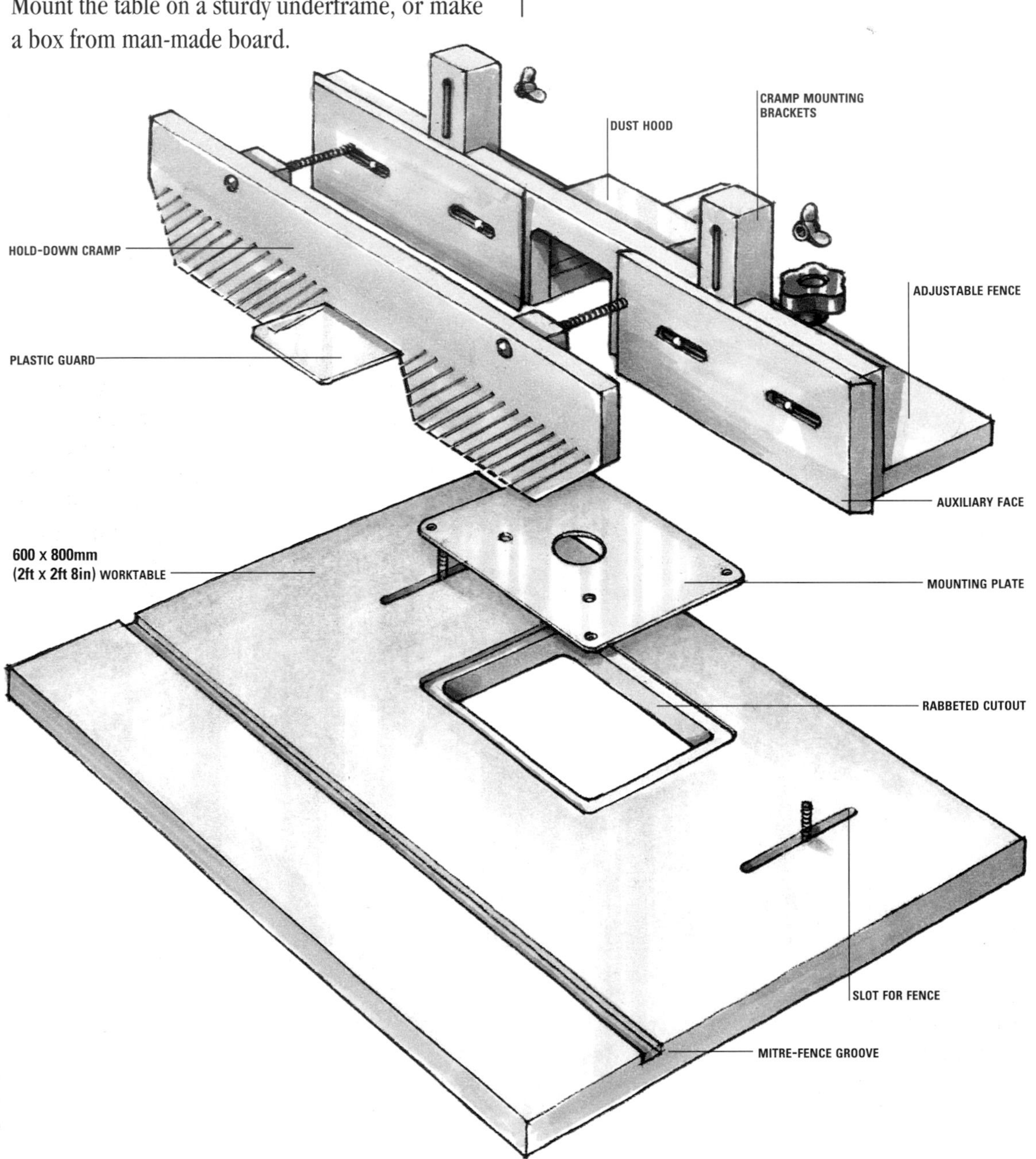

1 Cutting the table to size

Cut a 600 x 800mm (2ft x 2ft 8in) panel from 38mm (1½in) thick laminate-faced chipboard. Check that all corners are square and that the sides are parallel.

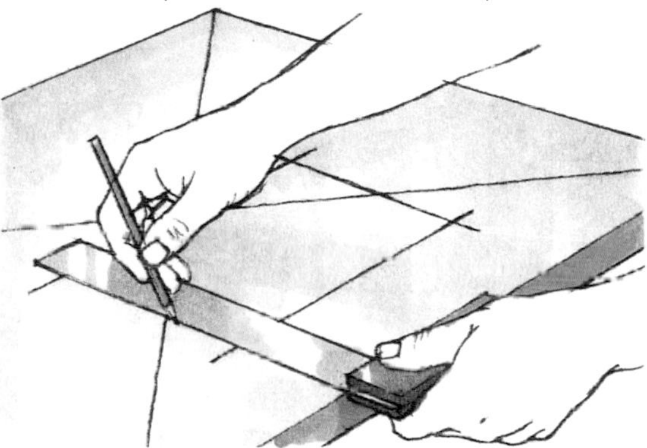

2 Marking out the panel

Mark the centre of the cutter position on the underside of the board. This can be dead centre on the board or offset to one edge. Mark out a rectangular cutout to accommodate the router.

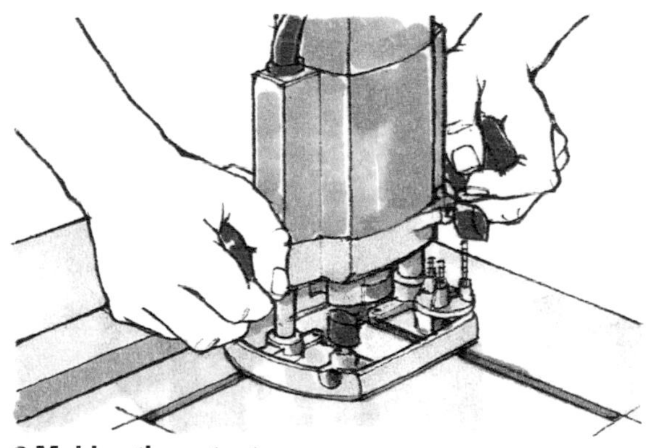

3 Making the cutout

Clamp the board flat on the bench with a piece of thin waste material between it and the bench top. Set up a router fitted with a TCT cutter, at least 10mm (⅜in) in diameter and long enough to cut through the board. Using a straightedge guide (see page 47), cut along each side of the rectangle in turn, working in stages until the board is cut through.

4 Machining the mounting-plate rabbet

Turn the board over and use a self-guiding rabbet cutter or a guide bush and template to machine a 12mm (½in) rabbet along each side of the cutout.

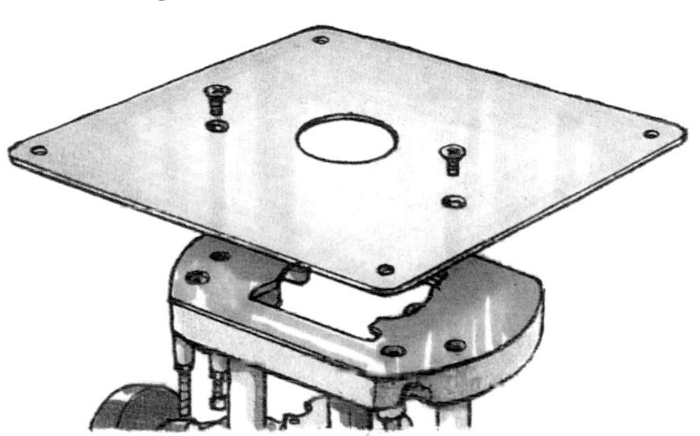

5 Cutting the mounting plate

Cut a rectangular plate from 3mm (⅛in) aluminium or steel, to fit into the table recess. Trace off a pattern from the router's base plate. Position the pattern centrally over the mounting plate and mark the centre of each mounting point and the cutter orifice. Drill and countersink the holes for the mounting screws and cut the orifice with a metal-cutting hole saw. A 44mm (1¾in) hole should be adequate in most cases.

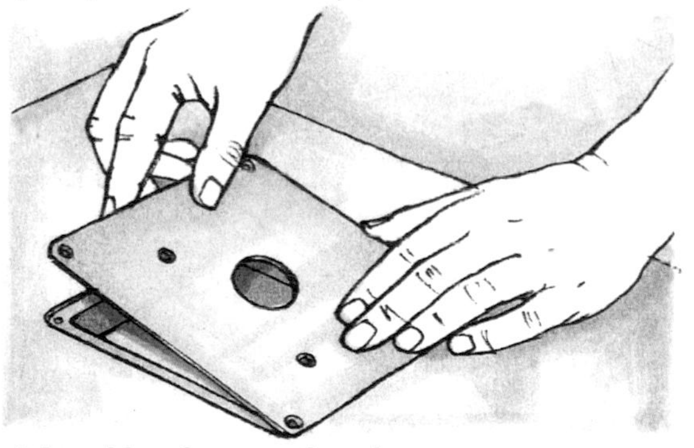

6 Attaching the mounting plate

Drill a 3mm (⅛in) hole in each corner of the plate and countersink them to take 50mm (2in) long mounting screws. Drill matching holes through the rabbeted edge of the table and secure the plate in the rabbet. Check that the plate lies flush with the surface, using thin shims or washers to level it.

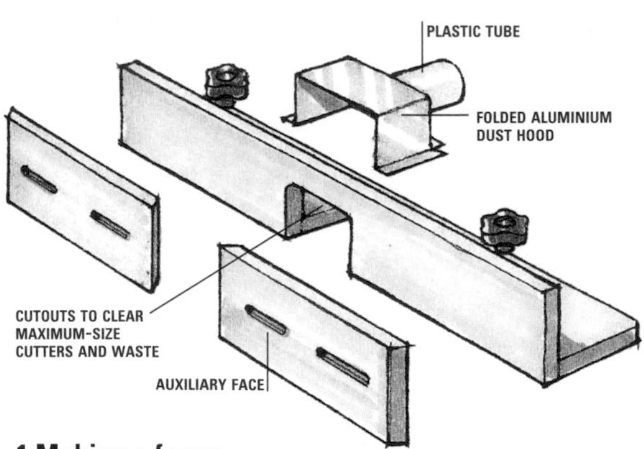

1 Making a fence

Plane two pieces of 75 x 12mm (3 x ½in) MDF square and parallel, then shape the cutouts. Glue and screw the wood together at an exact right angle. Make up a dust hood from thin plywood or folded aluminium and, using an epoxy-resin adhesive, glue a short plastic tube for the dust-extraction hose into an orifice cut with a hole saw.

2 Mounting the fence

Attach the fence, using two end-socket cramps (see page 40). Alternatively, cut slots through the table to take 6mm (¼in) coach bolts (the square shanks prevent them from turning). Cut recesses in the underside for the bolt heads.

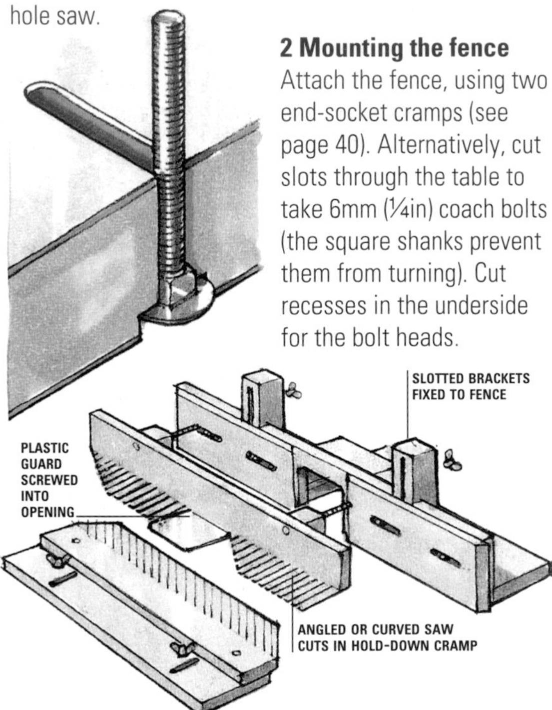

3 Making hold-down cramps

You can purchase hold-down or spring-pressure cramps for fitting to a table or fence. Alternatively, you can manufacture your own hold-down cramps by making a series of saw cuts in the edge of a piece of MDF or plywood. Fasten the vertical cramp to two slotted support brackets bolted to the side fence. Fit a small plastic top guard for additional protection.

A similar horizontal cramp fixed to a slotted batten holds the work against the fence. Attach it to the table, using coach bolts and wing nuts, or cramps.

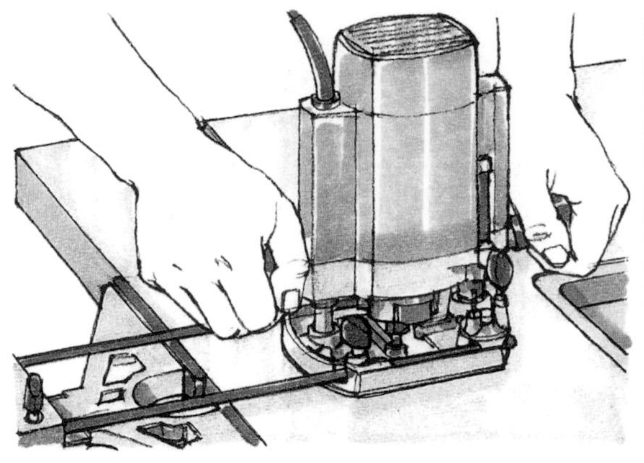

Cutting a slot for a mitre fence

To fit a proprietary mitre fence, measure the width of its narrow metal runner and machine a snug-fitting groove the length of the table and parallel to its edge.

USING LARGE-DIAMETER CUTTERS

To use large-diameter cutters that won't pass through the hole in a table mounting plate, line the surface with an MDF facing and fit the cutter from above.

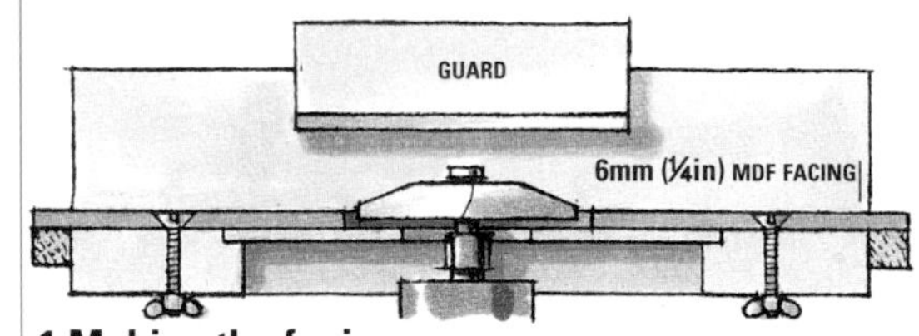

1 Making the facing

Lay a piece of 6mm (¼in) MDF over the table and, having marked the position of the cutter hole, cut a hole 3mm (⅛in) larger than the proposed cutter. Glue wooden battens along each side of the panel to hold it in position on the table. Drill and countersink two holes through the MDF and table, to take countersunk bolts and wing nuts.

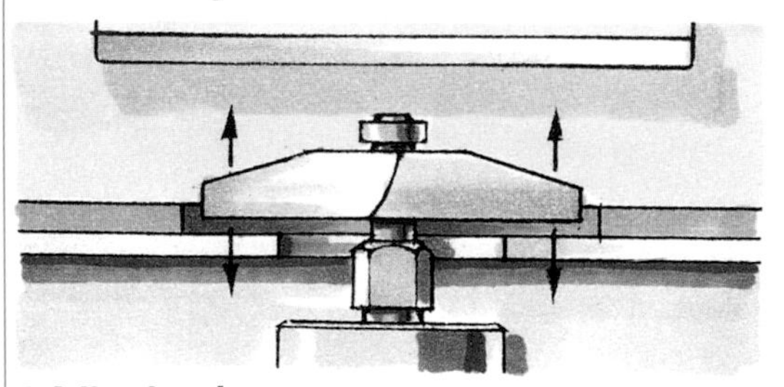

2 Adjusting the cutter

Using the fine-depth adjuster, raise the collet and fit the required cutter, passing the shank through the hole in the table. Make sure the shank is held securely (see page 28). Lower the cutter until its underside is fractionally below the surface of the MDF, but take care that it is not touching the table. Machine a trial piece to check the cutter setting.

WORKTABLE ACCESSORIES

Use a range of simple worktable accessories to help guide the wood safely past the cutter.

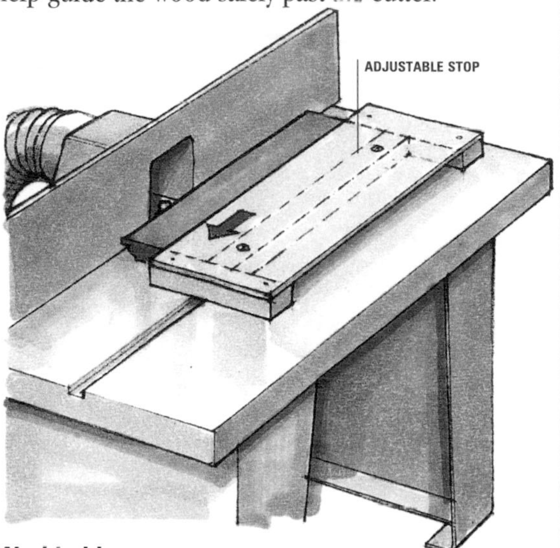

Workholders

Pass small workpieces safely past the cutter, using a purpose-made holder that keeps the wood pressed against the table and fence. Nail or screw a thin sheet of man-made board to battens that are exactly the same thickness as the work, and surround the workpiece on three sides.

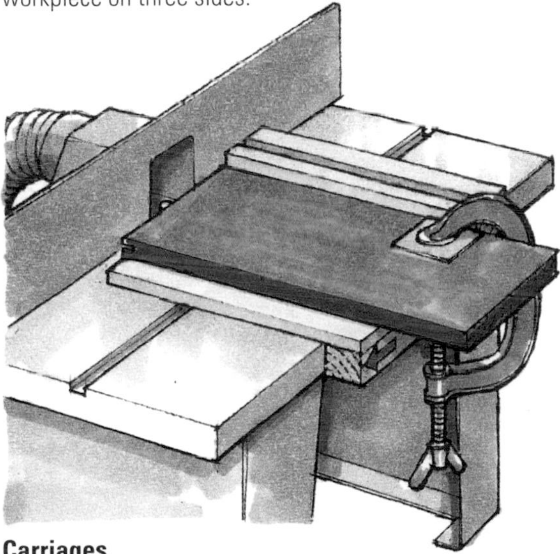

Carriages

Clamp a workpiece to a carriage that slides along the table. You can run a carriage against the edge of the table, or guide it using a slide batten running in the mitre-fence groove.

Fixed support

Clamped to the face of the table, a fixed support holds the work at a specific angle while it is fed against the table fence. To avoid having to run against a feather edge, don't remove too much timber.

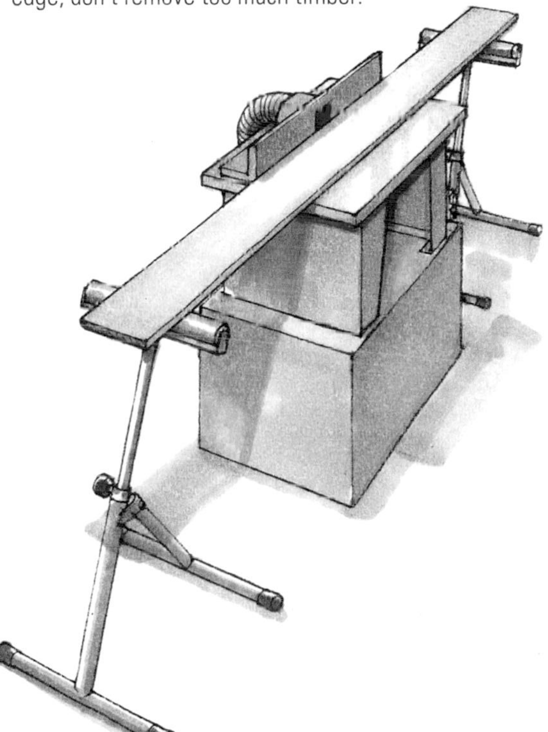

Roller stands

Always ensure that the workpiece cannot drop or tilt as it is fed into or as it leaves the cutter. For wide or long workpieces, provide extra support with a temporary side-extension table or use in-feed and out-feed roller stands at each end of the table.

WORKTABLE TECHNIQUES

For any table-routing operation, cut to the full depth in a series of shallow steps rather than in a single pass. The depth of each step will depend on the size of the cutter, shank diameter, cutter profile and router power (see pages 10, 11 and 20–6).

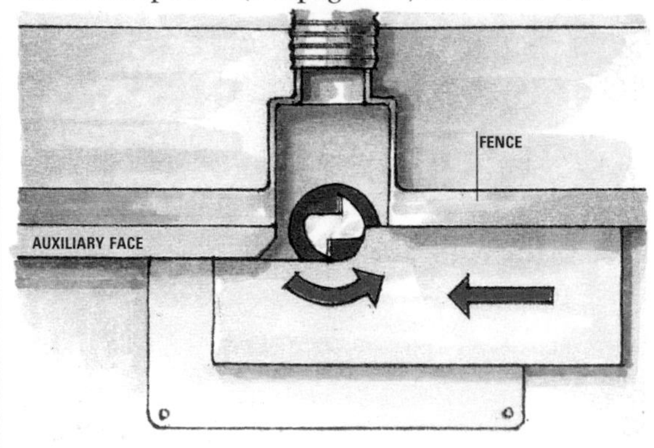

Feeding the work

When the router is inverted, the cutter rotates anti-clockwise. The feed direction must therefore be from the right of the table, against the rotation of the cutter. When machining the edge of a workpiece, set the fence to give the required width of cut, running the edge of the work against the fence face.

When shaping the full edge of a workpiece, fit a lead-out block or auxiliary face to the out-feed face of the fence. This compensates for the amount of material removed, and supports the workpiece as it is fed past the cutter.

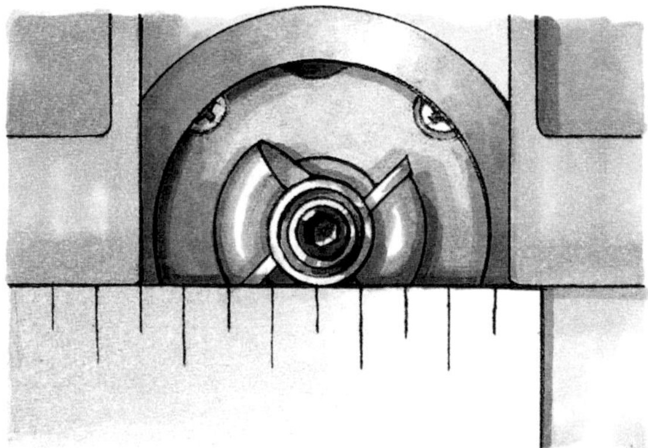

Using bearing-guided cutters

For straight work, set the bearing face flush with the fence, using a steel rule to align them. If the bearing is even slightly proud of the fence, it will deflect the workpiece at each end of the pass.

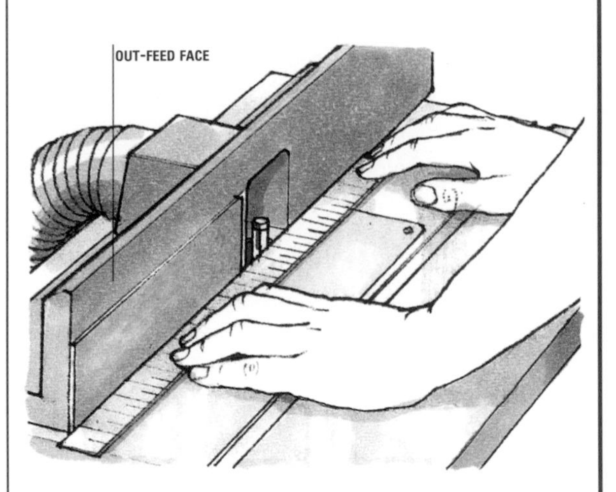

Edge planing

When planing and trimming edges, fit a 1.5mm (1/16in) thick auxiliary face to the out-feed face of the table fence – this is the most that should be taken off across the full depth of an edge in one pass. Fix the face to the fence with countersunk screws or double-sided tape.

Fit a two-flute cutter with cutting edges slightly longer than the thickness of the edge. Set one cutting edge against a steel rule held against the auxiliary face, making it proud of the in-feed face.

Hold the work firmly against the in-feed face until about half the workpiece has passed the cutter, then transfer the pressure to the out-feed face.

Rabbeting edges

When cutting a rabbet along the edge, hold the workpiece vertically against the fence or lay it flat on the table, with the rabbet edge running against the fence. Select a straight cutter equal to, or slightly larger in diameter than, the width of the rabbet.

Adjust the fence to make the required rabbet width, and raise the cutter to a suitable height for making the initial cut 3 to 4mm (1/8 to 5/32in) deep. Machine to the full rabbet depth in stages, raising the cutter by similar amounts between passes.

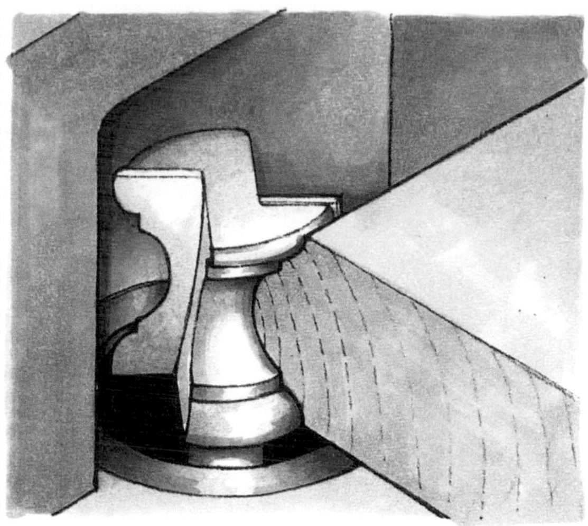

Cutting decorative mouldings

Machine decorative mouldings in a similar fashion to rabbets, with the workpiece held vertical against the fence or flat on the table. By adjusting the fence and altering the height of the cutter, you can use all or part of the cutter profile to make different-shape mouldings.

You can also alter the profile by presenting the work at different angles to the cutter. Use a fixed support (see page 75) to maintain the required angle throughout the pass.

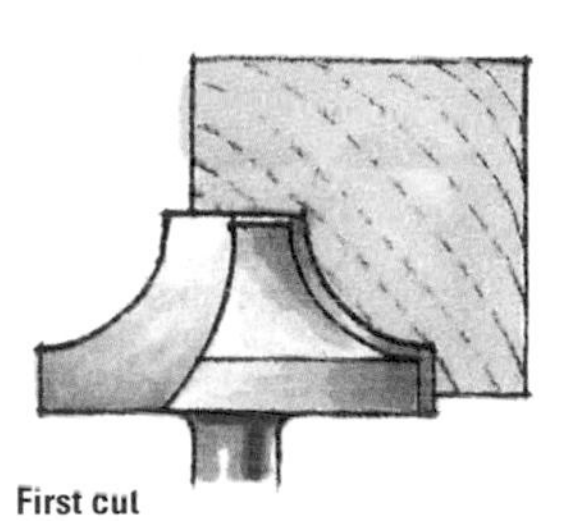

First cut

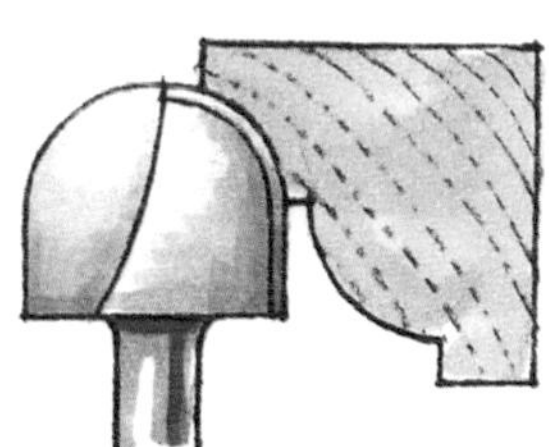

Second cut

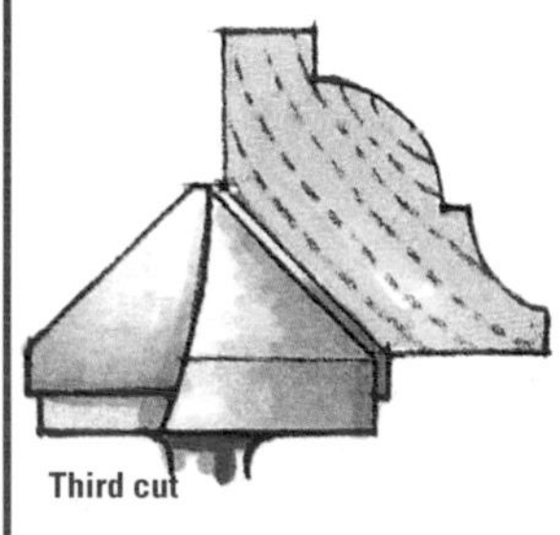

Third cut

Cutting composite mouldings

It is not always necessary to buy a specific cutter in order to machine a particular moulding, as it is quite possible to make use of a combination of cutters to make up the required composite profile.

When cutting composite mouldings, always work out in advance the most practical sequence for cutting each part of the profile. Use the flat faces of the workpiece as guide edges before they are cut away to an inadequate edge that may need to be supported by a scribed face on the out-feed side of the fence.

CURVED AND SHAPED WORK

Curved and irregular-shape work can be machined on the inverted table, using bearing-guided cutters or by mounting a bearing directly above a plain cutter. Whichever method you use, fit a guard. Take care when handling small or narrow workpieces. Always make up a simple workholder to provide better and safer control of the work.

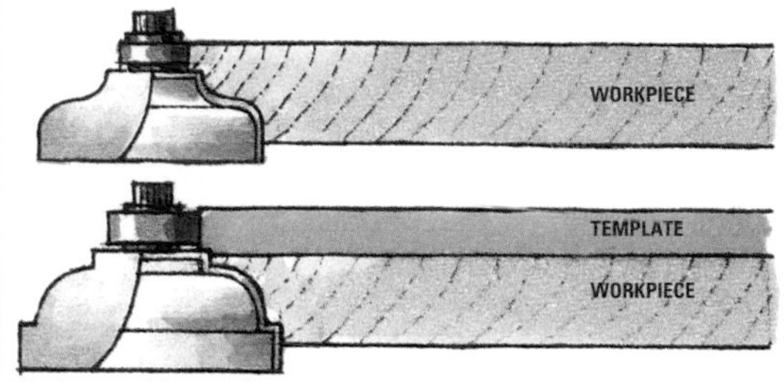

Using self-guiding cutters

When edge-moulding, run the work against the guide bearing, or guide the work by means of a template attached to the wood (see page 56). Always hold the work firmly against the bearing and lead-in block.

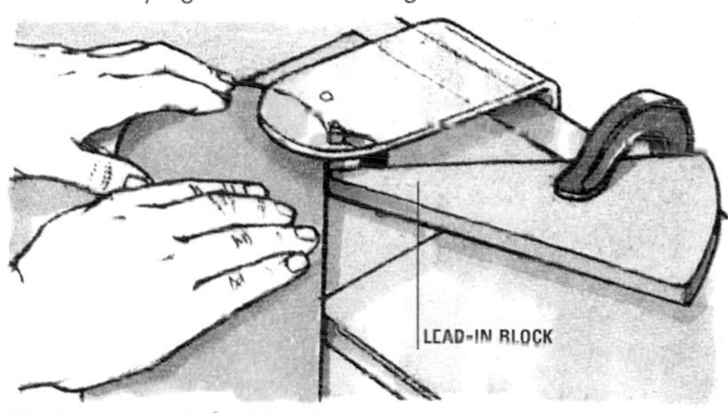

Fitting a lead-in block

To minimize the risk of the cutter snatching the workpiece as it is fed across the table, use a home-made lead-in block clamped to the surface. A similar block can be used to prevent the work being pulled behind the cutter as it leaves. However, this may not be possible when machining small-radius curves.

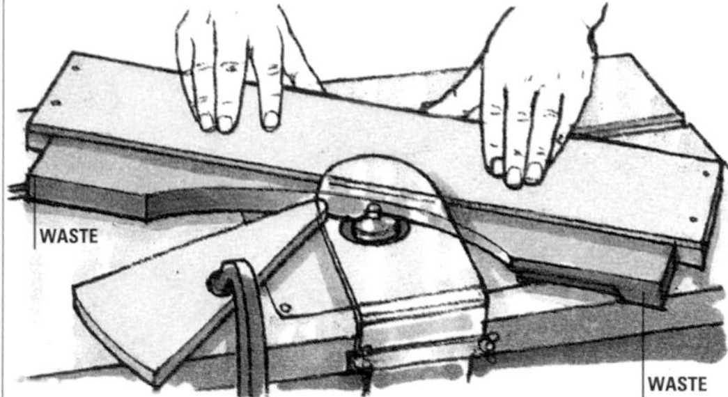

Lead-in and lead-out waste

Allow extra waste at both ends of the workpiece when trimming or moulding curved or shaped edges. This provides sufficient material to allow the work to be fed smoothly onto and away from the cutter.

GROOVES AND HOUSINGS

Cut grooves and housings with the face edge or a square end of the workpiece against the fence. Machine housings across a narrow board using the table's mitre fence. Since it can be difficult to guard the cutter when machining housings in very wide boards, it is safer to run a hand-held router against a straightedge guide (see page 47), or use an overhead-router table (see pages 80–4).

Using the table fence
Set the cutter height and make the first pass, keeping the face of the work flat on the table and the guide edge held firmly against the fence. To avoid break out on the rear edge when cutting a housing, hold or clamp a batten against the rear edge and pass it across the cutter with the work.

Using the mitre fence
For standard housings, set the sliding mitre fence at 90 degrees to the table fence. Sandwich a waste batten between the work and mitre fence then, holding the work firmly in place, feed the fence towards the cutter. Adjust the mitre fence to make angled housings.

Cutting dovetail housings
You can cut a dovetail housing with a hand-held router or as described left. To cut a matching tail on the end of a board, adjust the cutter height and the fence to match the dimensions of the housing, then machine a trial piece to check your settings.

Holding the work vertically against the table fence, feed it past the cutter. Turn the piece around and cut the other face. Check that the tail is a snug fit in the housing and make adjustments as necessary before cutting the actual workpiece.

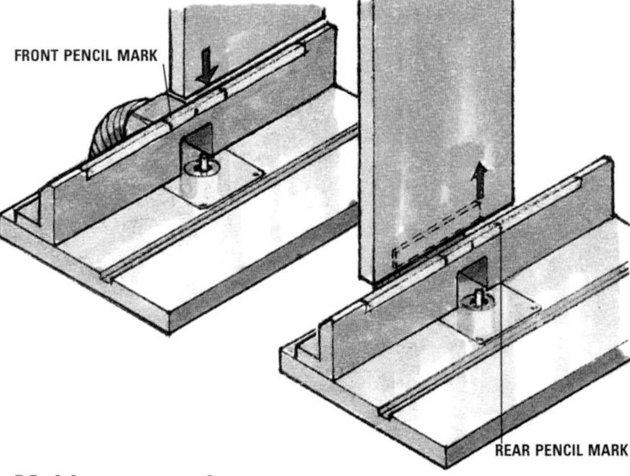

Making stopped cuts
As the cutter in an inverted router cannot be plunged into the surface of the work, you have to make stopped grooves, chamfers and mortises by lowering the wood onto the cutter. Stick a piece of masking tape along the top edge of the fence, and mark in pencil the positions of the stopped ends before and behind the cutter. Holding the front edge of the workpiece against the front pencil mark, lower the wood onto the cutter. Feed the wood forwards until the back edge of the work aligns with the rear pencil mark, then back off slightly and carefully lift the work clear of the cutter.

USING JOINTING CUTTERS

An inverted router is essential when using the special cutter sets employed to make finger joints, comb joints and scribed tongue-and-groove joints. You can usually machine the full-width edge of the workpiece, but adjust the table fence between passes to machine to the full depth in a series of shallow cuts. Before cutting the actual workpiece, always machine trial pieces to check that the joint fits satisfactorily.

Stile-and-rail cutter

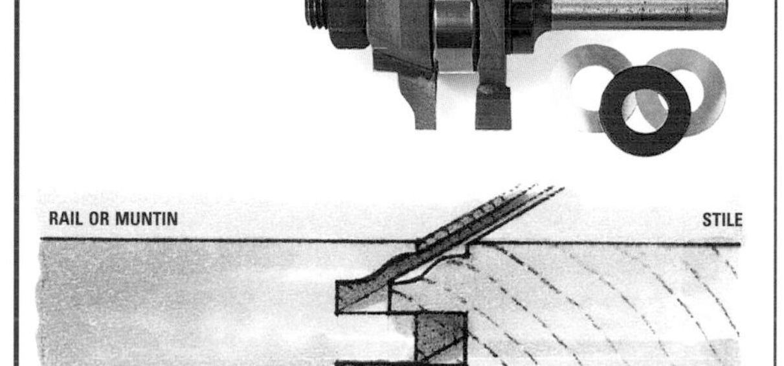

Making moulded frames

You can joint, groove and mould panel-door frames using a single set of stile-and-rail cutters. In one arrangement, the set will cut a groove and moulding along the inside of each frame member; when rearranged, the same set of cutters will produce a scribed moulding and tenon on the end of the rails or muntins. A cutter set with a 6 or 10mm (¼ or ⅜in) shank can be used with a medium-duty inverted router.

When cutting rail or muntin ends, use a carriage to keep the workpiece square to the table fence, and use a waste batten to prevent break out on the back of the timber (see page 75).

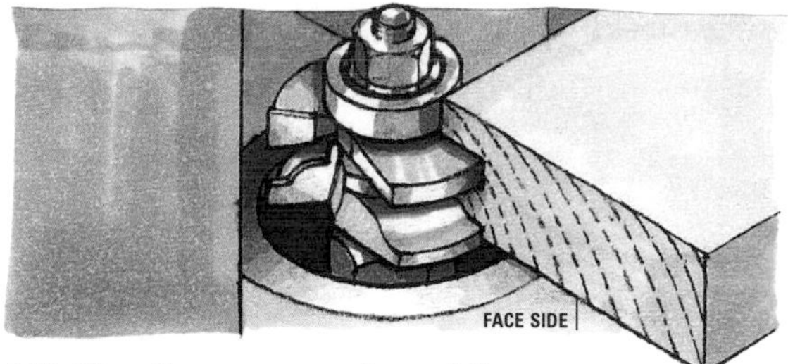

1 Cutting the groove and moulding

Assemble the cutter set and adjust the height to suit the thickness of the timber. Set the table fence so that no more than one third of the cutter profile projects, and make one pass. Re-set the fence between passes until you have cut to the full depth.

CUTTING LAPS AND TENONS

Cut laps and tenons using the mitre fence or a home-made sliding carriage (see page 75). You can cut several narrow rails at the same time. Always place a batten behind the work to prevent break out.

2 Cutting the scribed moulding and tenon

Unplug the router and rearrange the cutter components for the reverse scribing, and match it against the cut stile. There is no need to reset cutter height. Cut the moulding and tenon on a trial rail, and make adjustments until a perfect joint is formed.

When cutting the actual workpiece, cut the scribed moulding and tenon first, since any break out will be removed when the edge moulding and groove are cut.

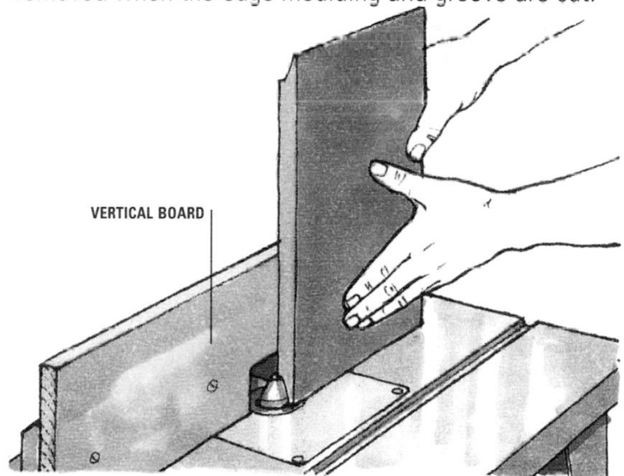

Machining a raised panel

You can machine the edge of a door panel using a vertical panel-raising cutter, suitable for a medium-duty router. The relatively thin panel must be kept vertical throughout the pass, so fix a deep face to the fence. Machine the edges in stages, raising the cutter between each pass. Make the first cuts across the grain at each end of the panel, since break out will be removed when the side edges are machined.

OVERHEAD-ROUTER STANDS

Some woodworkers prefer to mount a fixed router above the worktable. The advantages are that the actual cutting action is clearly visible and waste material can escape easily, rather than being trapped beneath the work or between the work and the fence. Professional overhead stands are fitted with a foot-operated plunge action, which allows you to control the workpiece with both hands throughout an operation. Most overhead stands for home workshops are fitted with a hand-operated plunge action (similar to a drill press) or even a simple height adjuster. Most operations carried out on these stands are fed from the edge, rather than being plunge-cut.

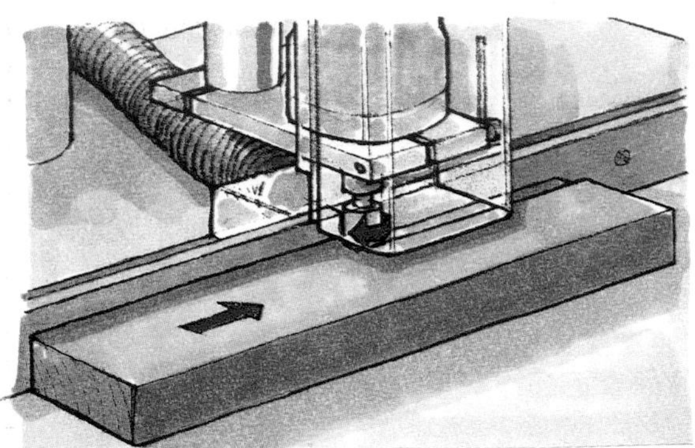

Feed direction
The cutter in an overhead router rotates clockwise, so the work must be fed into it from left to right.

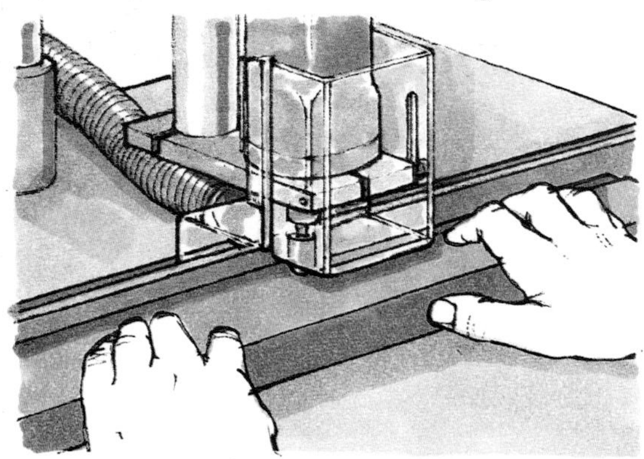

Fences and guards
A simple straight fence can be clamped across the table. The cutter, mounted above the work, is guarded by the router's base plate. However, a separate guard must be fitted to light-duty routers that can be separated from their plunge carriage and base plate.

Proprietary router stand

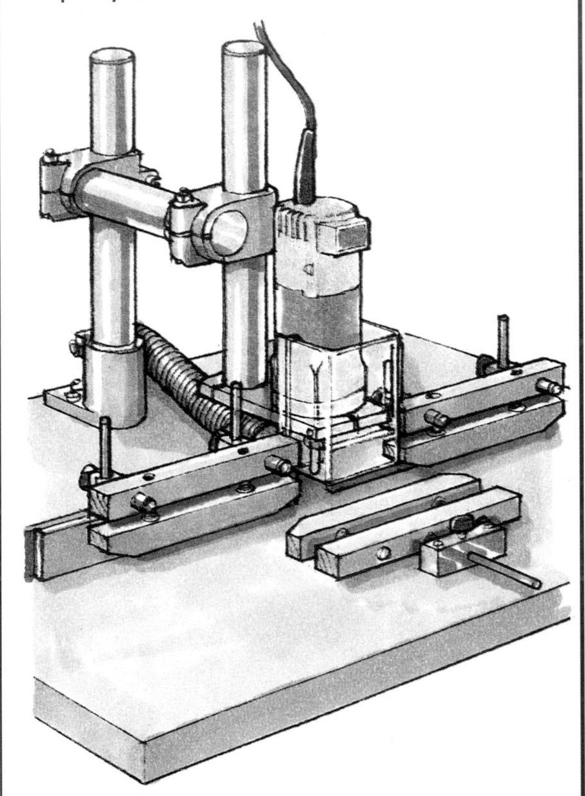

Hold-down cramps
You can fit a horizontal side-pressure cramp to hold the material against the fence. You may need to position vertical hold-down cramps on each side of the router.

MAKING AN OVERHEAD STAND

You can make a simple fixed overhead stand, using a wooden block drilled to take the router's side-fence rods. Clamp the finished stand firmly to a work surface.

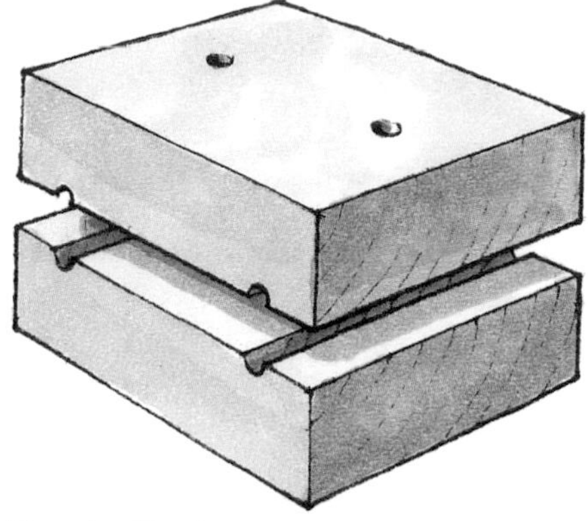

1 Making the block
Cut a hardwood block, 125 x 100 x 75mm (5 x 4 x 3in). Check that all sides are square and parallel. Drill two holes to accommodate the router side-fence rods. Cut the block in two through the centres of the drilled holes. Drill two holes vertically through the blocks to take 8mm (5⁄16in) mounting bolts.

If you are not equipped with a suitable saw, clamp two separate blocks together and drill the holes for the fence rods where they join.

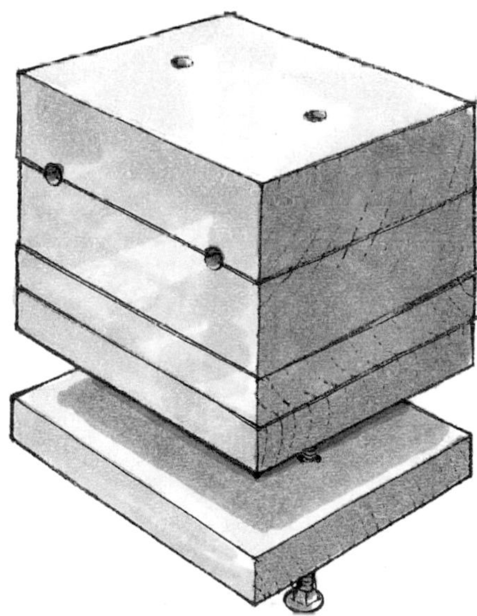

2 Making spacer blocks
Cut spacers the same width and length as the mounting blocks, and drill matching bolt holes. These blocks are used to adjust the height of the router.

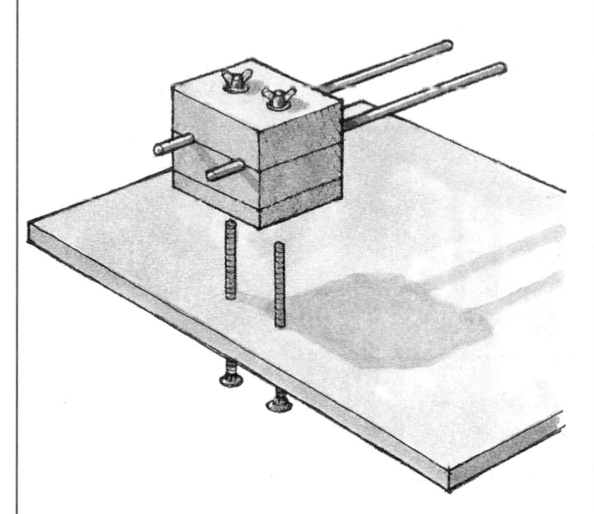

3 Mounting the blocks on a base board
Cut a 600 x 500mm (2ft x 1ft 8in) base board from 18mm (3⁄4in) MDF or a similar material. Drill two holes close to one edge for the mounting bolts. Fit the fence rods and bolt the block and spacers to the base board, using washers and wing nuts.

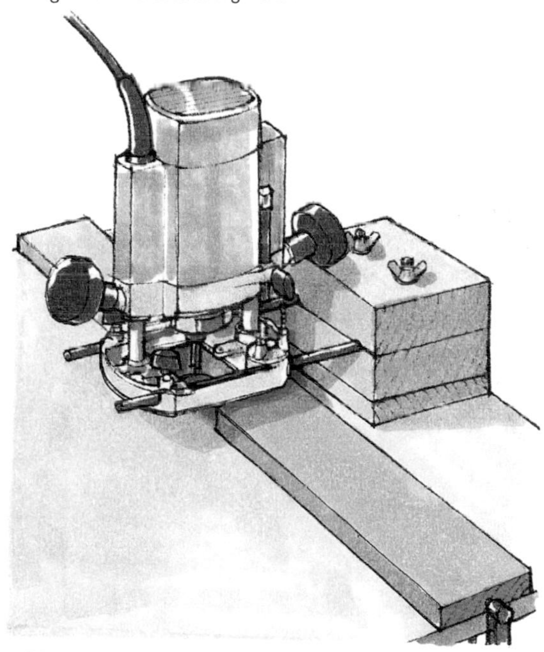

4 Making fences
An assortment of simple fences can be used to accommodate different cutters. Cut a fence from 18mm (3⁄4in) MDF or hardwood, and form a semi-circular cut-out to accommodate the cutter. Alternatively, make a fence similar to that described for an inverted router (see page 74).

RADIAL-ARM ROUTING

Many home workshops are equipped with a radial-arm saw. It is a compact and versatile machine, used primarily for crosscutting and ripping timber and boards to size, but it can also be adapted to drill, sand and shape workpieces. In addition, replacing the saw blade and guard with a router-mounting bracket enables a radial-arm saw to be used as a versatile overhead-router table.

You can cut standard housings, grooves and rabbets using the saw's crosscut and ripping facilities, while swivelling and tilting the router presents the cutter to the work at various angles. To make stopped cuts, utilize the router's plunge mechanism or crank the radial arm up and down.

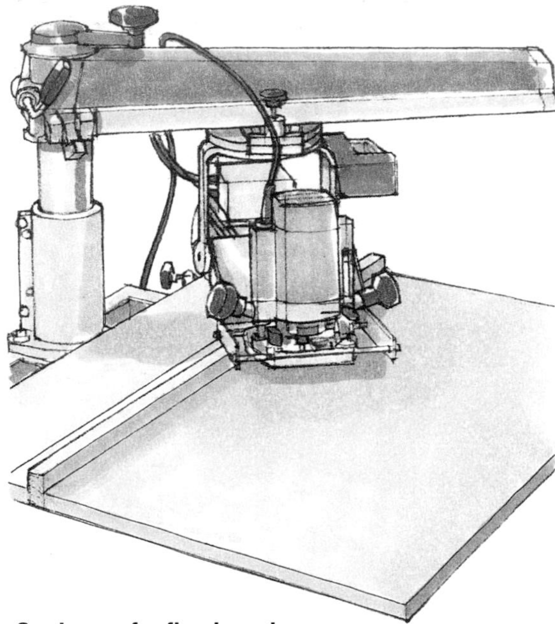

Setting up for fixed routing
Start by disconnecting the saw's motor from the power supply; for safety, fit a no-volt release switch beneath the front edge of the table (see page 71).

Swinging the arm at an angle provides better access to the router controls and gives you a clear view of the cutter when feeding a workpiece along the fence.

Line the worktable with a thin sheet of waste board so that you can cut right through a workpiece without damaging the table surface. Pin the sheet at the edges where the router cutter cannot reach. A fixed base plate will serve as a cutter guard, but fit some other form of protection if the router motor can be detached from its carriage and base plate.

USING THE TILTING YOKE

Use the tilting yoke to set the cutter at an angle to the workpiece. Fit straight cutters to router V-grooves, and set shaped moulding cutters at different angles to alter moulding profiles (see pages 24–6).

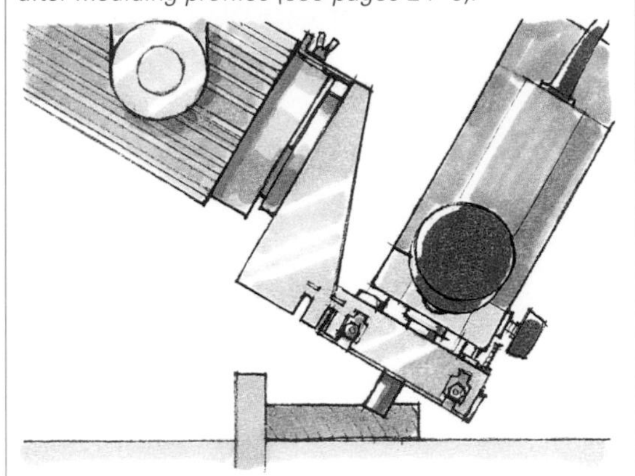

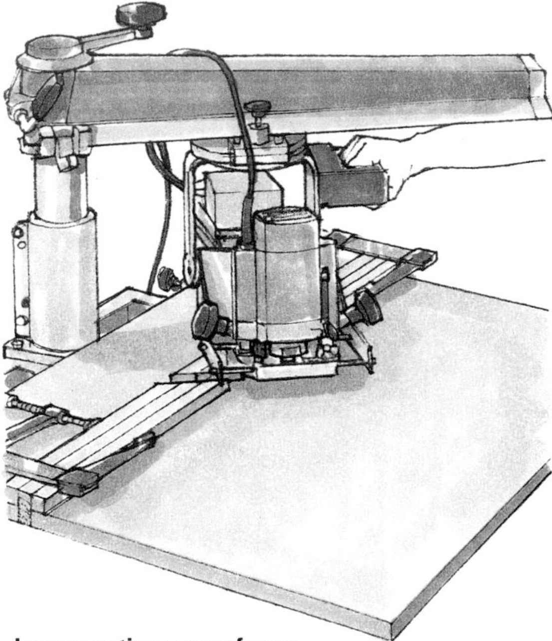

Incorporating a rear fence
Replace the rear table fence with a new strip of similar-size wood or board, held in place with the table's variable cramps. Cut a notch in the replacement so that you can position the cutter within the line of the fence for edge-moulding operations. Fit a straight cutter that is larger than other cutters you intend using, allowing 3 to 4mm (⅛ to 5/32in) clearance on either side.

Clamp a batten on each side of the fence, to prevent break out, then lower the router cutter to the table surface. Release the radial-arm yoke and, with the router running, feed the cutter through the fence.

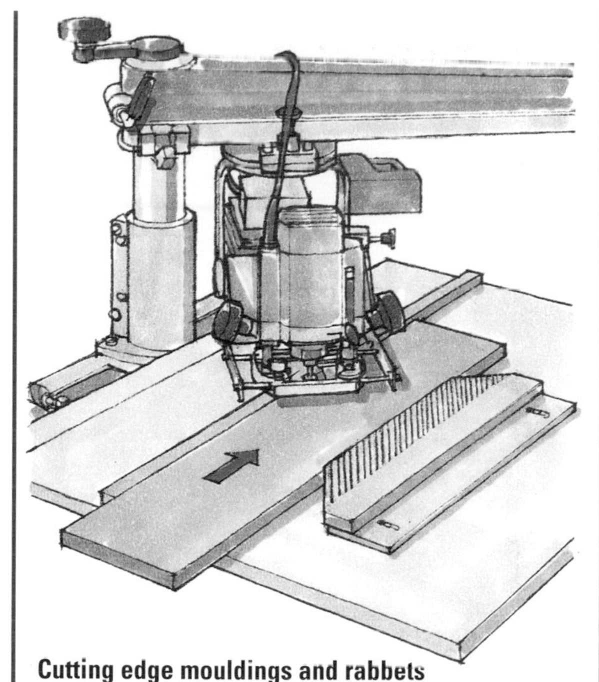

Cutting edge mouldings and rabbets
Set the cutter height, and position and lock the yoke for making the first pass. Hold the workpiece firmly against the fence throughout the pass and work from left to right, against the rotation of the cutter. For safety, set up a side-pressure cramp against the front edge of the workpiece (see page 74).

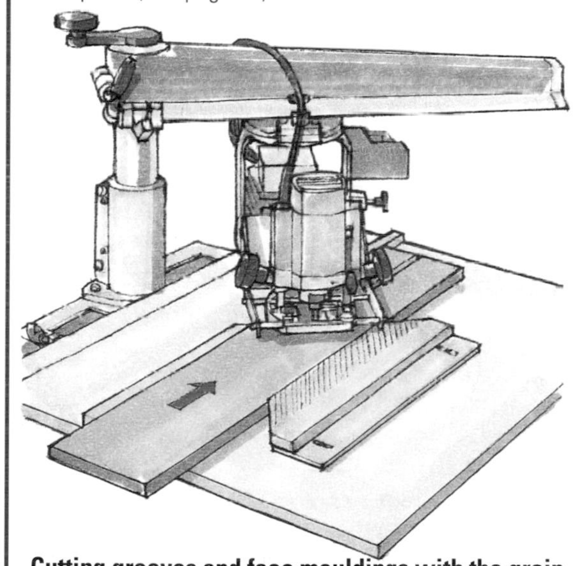

Cutting grooves and face mouldings with the grain
Slide the router along the arm until the cutter is positioned the required distance from the edge of the workpiece, then lock it in place. Set the cutter height and make the first pass as described above. Machine deep grooves in stages, adjusting the depth of cut between passes.

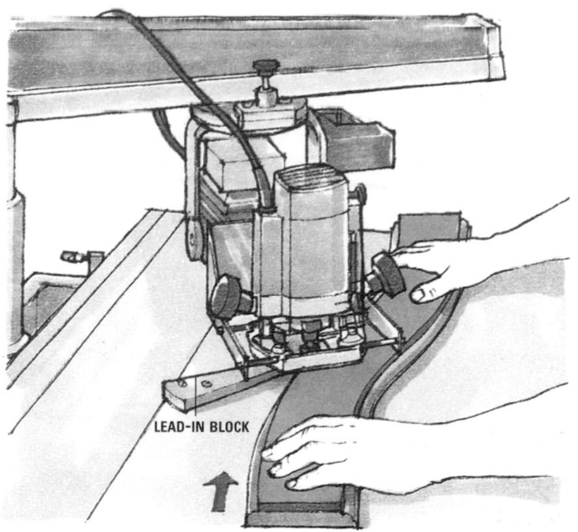

Cutting external curves and shaped edges
For moulding and trimming, remove the rear fence, position the router in the centre of the table, and fit the required self-guiding cutter. To prevent the work from being snatched away by the cutter, fix a lead-in block to support the work.

Moulding and trimming internal shapes and curves
Cut a template to the required size and shape for use with a bottom-cutting, self-guiding cutter. Mark out the aperture on the workpiece and, using a jigsaw, cut out the waste to about 2mm (3/32in) inside the line.

Tape or pin the template to the underside of the workpiece. Clamp the assembly temporarily to the worktable and lower the router to the required level. Switch on the router and release the cramp; then, holding the work firmly in both hands, feed it against the cutter in an anticlockwise direction to make the cut in one pass. When cutting large profiles, lower the router in stages.

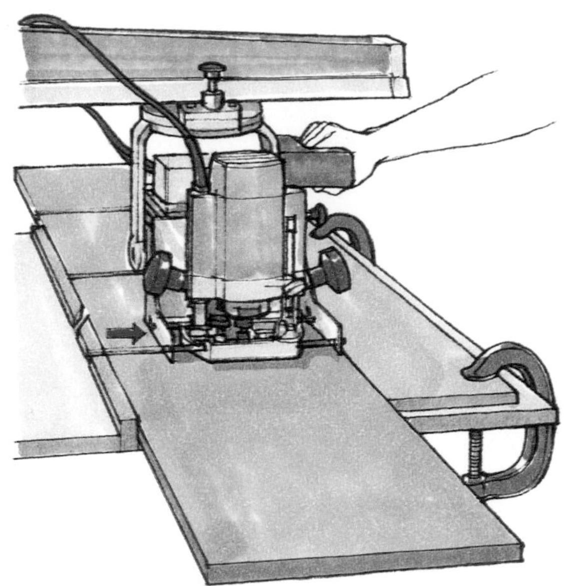

Cutting a housing

Cutting housings across the work requires a different set-up. In this arrangement, the radial arm is set at the required angle and the router is moved across the workpiece, which is clamped against the fence, using a similar-thickness board. Set the depth of cut, then pull the router towards you to make the first pass. Lower the router behind the workpiece between each subsequent pass.

Cutting stopped housings and mortises

When machining stopped housings, attach a small G-cramp to the radial arm, to prevent the router moving further than required to make the cut. A second cramp behind the router will limit travel in both directions. Start by plunging the cutter into the work, then pull the router towards you until it reaches the stop cramp.

Don't let the router come to rest for too long, or friction will burn the wood.

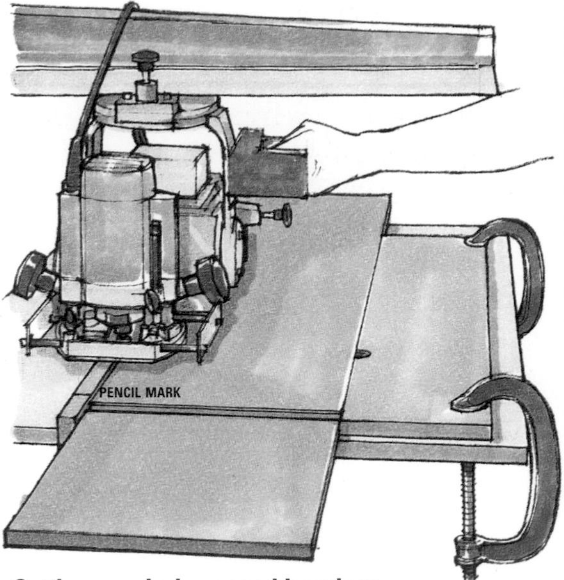

Cutting regularly spaced housings

To cut a series of housings or slots, place the workpiece against the fence and machine the first housing to the required depth. Mark the position of the next housing on the fence, then release the workpiece and slide it between the fence and cramp board until it aligns with the mark on the fence. Cut further housings in the same way. For angled housings and louvre mortises, swing the radial arm to the required angle.

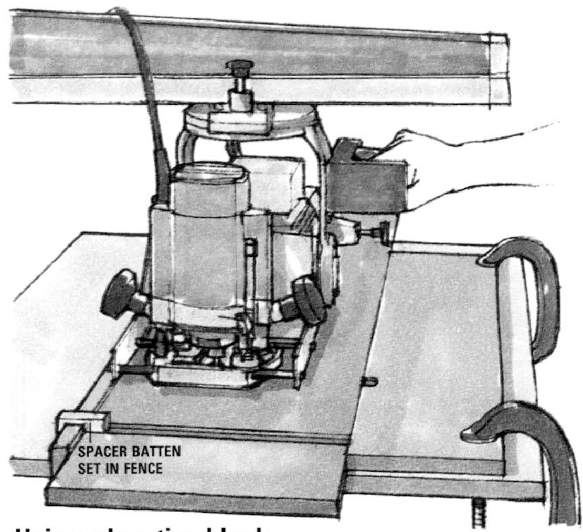

Using a locating block

An alternative method for cutting regularly spaced housings is to cut a notch in the fence equal to the size of the housing and insert a tight-fitting block of wood, to position the workpiece for each subsequent cut.

WORKCENTRES

There are proprietary workcentres designed for fixed-routing operations, some of which can also be fitted with a portable power saw for ripping and crosscutting timber. These workcentres can be set up as an inverted-router table, or the router can be fitted to a sliding carriage that passes over the work. On some workcentres, the tool is mounted overhead and the workpiece is clamped to a sliding carriage beneath it. Using any of these methods, you can control the cutter through the work more precisely than with a hand-held router.

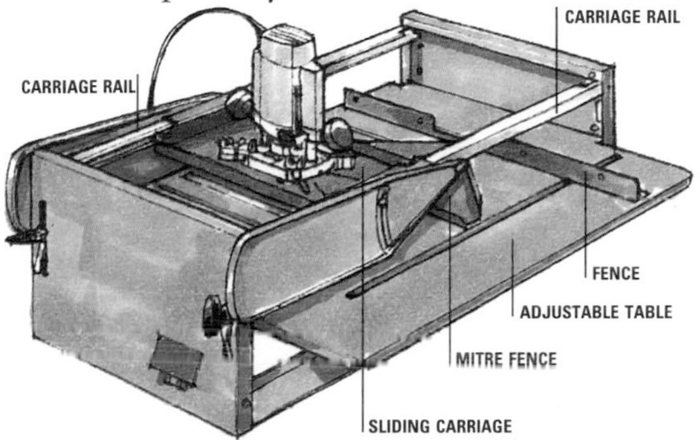

Setting up a sliding carriage
Set up the adjustable-height table beneath and parallel to the carriage rails. Brace the workpiece against a two-piece fence to cut at 90 degrees. For other angled cuts, clamp the workpiece to an adjustable mitre fence.

Routing on a sliding carriage
The carriage is mounted on twin parallel rails above a fixed table, and is mainly used for cutting edge rabbets, mouldings and housings across the grain, as well as machining short lengths of wood along the grain. Most makes of router can be clamped or bolted to the carriage. A flip-over facility and removable table mean you can invert the router when required.

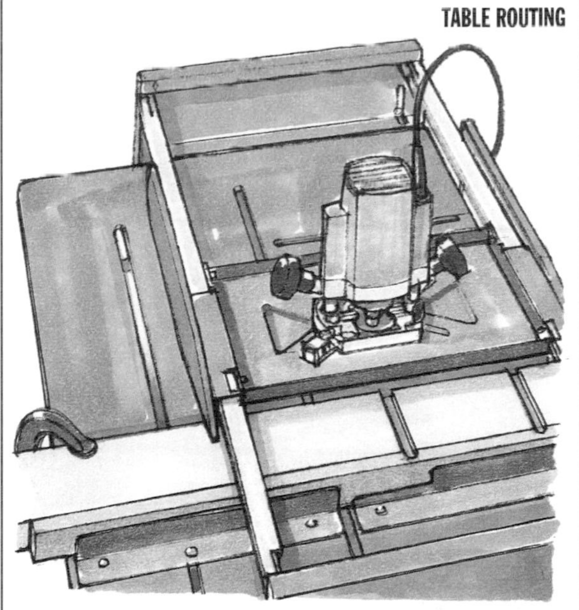

Making crosscuts
Insert a batten of the same thickness as the workpiece between the fence and rear edge of the material. This prevents break out as the cutter emerges from the work. Clamp the workpiece securely across the table.

Mark out the position and depth of the cut on the workpiece. Set the depth of the first cut on the router's turret stop, checking the position against the marked workpiece. Machine towards the fence, then raise the cutter and bring back the carriage before making the next pass.

Making stopped cuts
To cut stopped housings and mortises, attach a small cramp to the carriage rails, limiting the distance the router can travel.

SLIDING-CARRIAGE WORKCENTRE

In this system, the router is mounted overhead on a sliding plate, which is precisely positioned between guides and locked to a rigid plate. The router can be fixed at any point, and the workpiece is then wound from side to side or pushed into the cutter. Alternatively, the workpiece can be clamped beneath the router plate, and the router passed back and forth. These methods perform operations such as joints, grooves, slots, recesses, profiles and mouldings.

Feeding the workpiece
For machining the ends of a workpiece, a cam-operated cramp holds the work vertically to the face of the sliding carriage. The carriage is wound sideways, using a hand wheel, which feeds the wood into the cutter. Feed direction can be either against or with cutter rotation; cutting with the rotation virtually eliminates break out and leaves neat, clean shoulders.

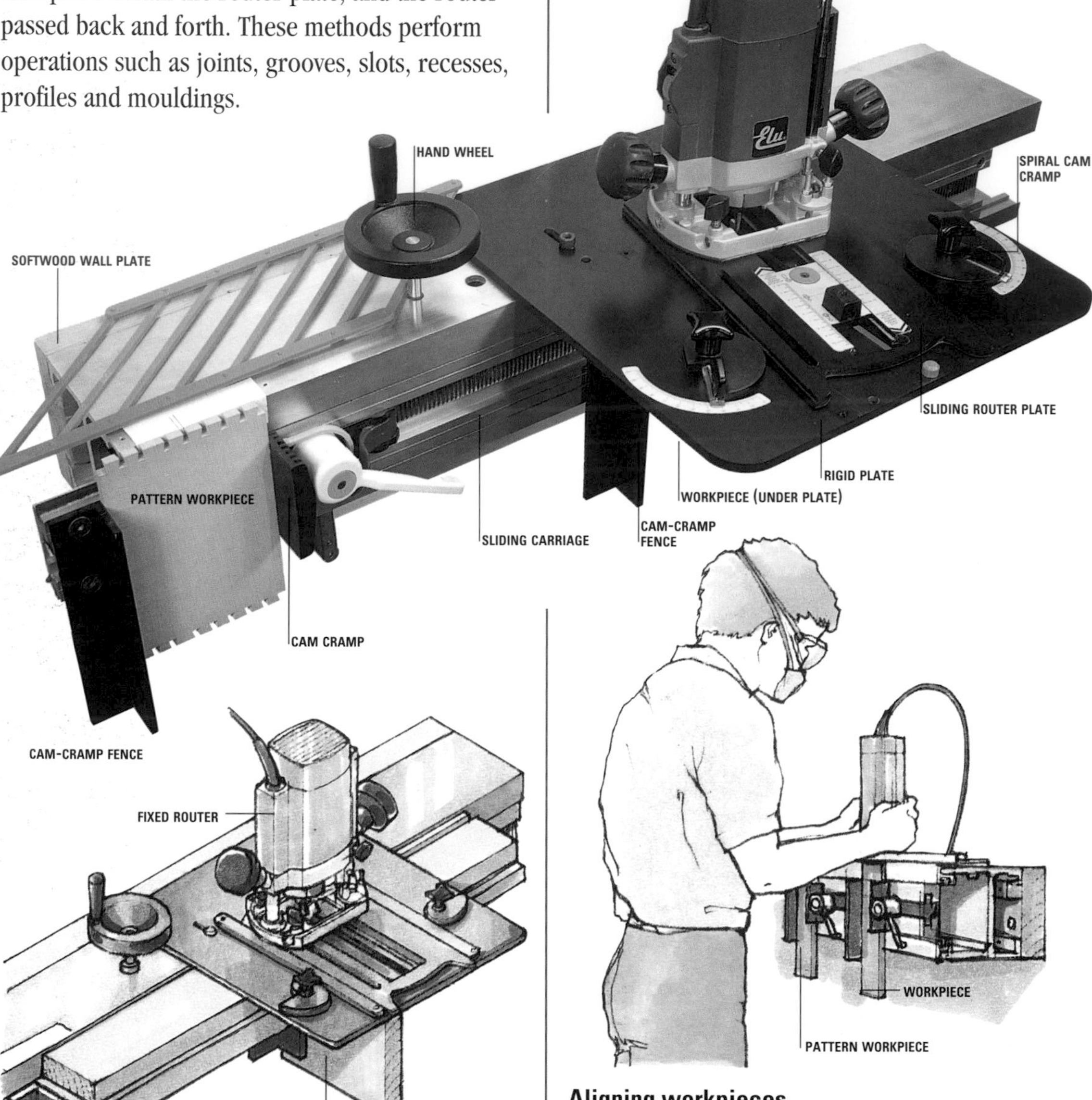

Clamping long workpieces
The workcentre cramps can be adapted to hold or support long workpieces for grooving and rabbeting.

Aligning workpieces
A unique feature of this system is a second cramp on the sliding carriage to accommodate a pattern, which allows workpieces to be quickly and accurately aligned with the cutter. A number of identical components can be clamped together and cut in one operation.

CHAPTER 8

It is possible to make simple joints, such as housings, rabbets and grooves, following basic procedures described earlier in this book. In addition, there are special woodworking joints for cabinetmaking and joinery, some of which can be quite laborious to construct by hand, but which you can make accurately and relatively quickly, using a power router with the appropriate jigs and guides.

HALVING JOINTS

The halving joint is used exclusively for framing that is made from components of equal thickness. Though it is not an especially strong joint, it is extremely simple to make, and the basic joint can be adapted to make right-angle corners, T-joints and cross frames. As with all halving joints, an equal amount of wood is cut from each workpiece so the components lie flush when the joints are assembled.

Corner halving joint

Mitred halving joint

Oblique halving joint

Cross halving joint

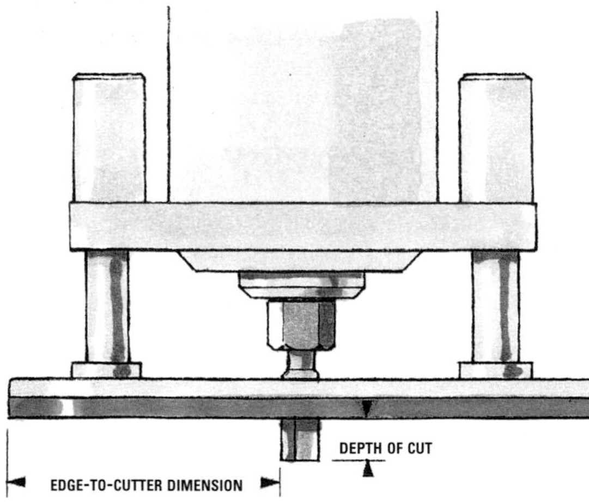

Setting up the router

Set the maximum depth of cut to half the thickness of the timber, and set the turret stop to cut in two or more passes. Measure from the edge of the base to the side of the cutter – use this edge-to-cutter dimension to position the jigs on the work.

MAKING RIGHT-ANGLE JOINTS

A pair of simple L-shape jigs will enable you to cut any of the halving joints with ease. Glue and screw together 300mm (12in) lengths of 75 x 18mm (3 x ¾in) and 150 x 18mm (6 x ¾in) prepared wood. You can cut oblique halving joints simply by making the jigs to the required angle.

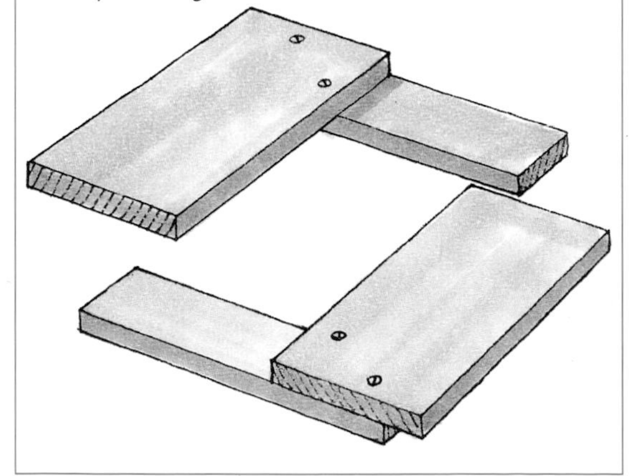

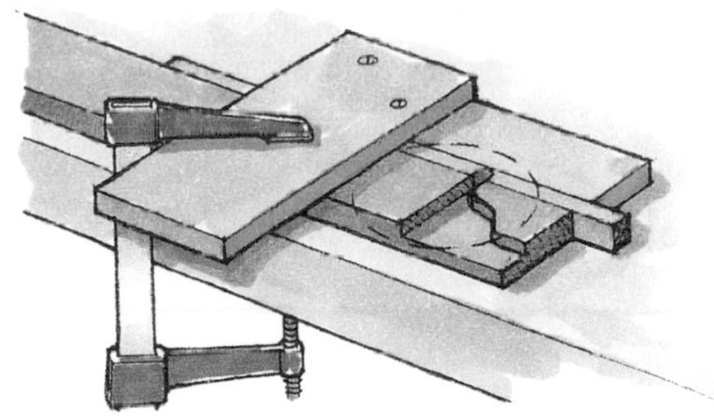

Marking and cutting a corner halving
Draw the shoulder line across the work and mark another line, the edge-to-cutter dimension (see opposite), from the shoulder line. Align an L-shape jig with this second line, and clamp both the jig and work to the bench. To cut the halving, run the edge of the router base against the jig to form the shoulder, then remove the rest of the waste freehand. The jig itself will prevent break out, but if you want to keep the jig intact, place a strip of scrap wood behind the work.

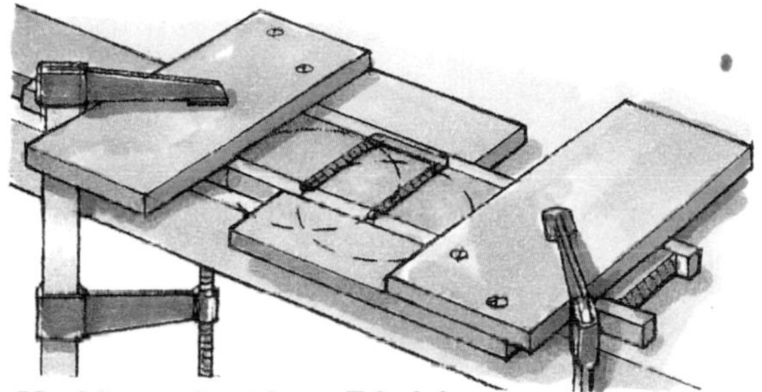

Marking and cutting a T-halving
One half of a T-halving joint is cut as described above. To cut the recess, mark two shoulder lines and clamp two jigs on top of the work, allowing for the edge-to-cutter dimension. Without altering the depth setting on the router, use the left-hand jig to cut one shoulder, and then move the router to the right-hand jig to cut the other. Clean out the waste in between with the router.

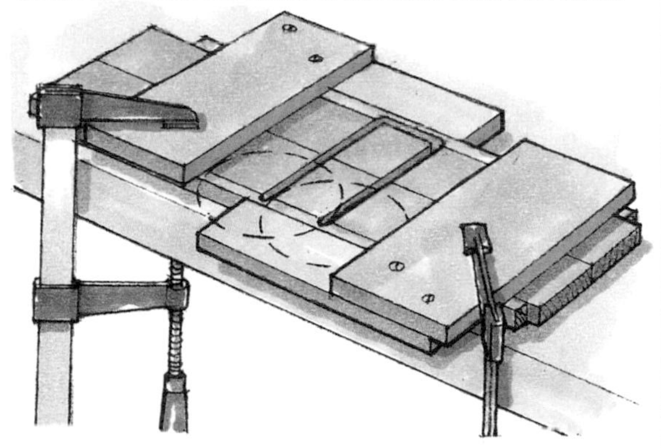

Making cross halving joints
Clamp the two components of a cross halving joint together, and cut both recesses simultaneously.

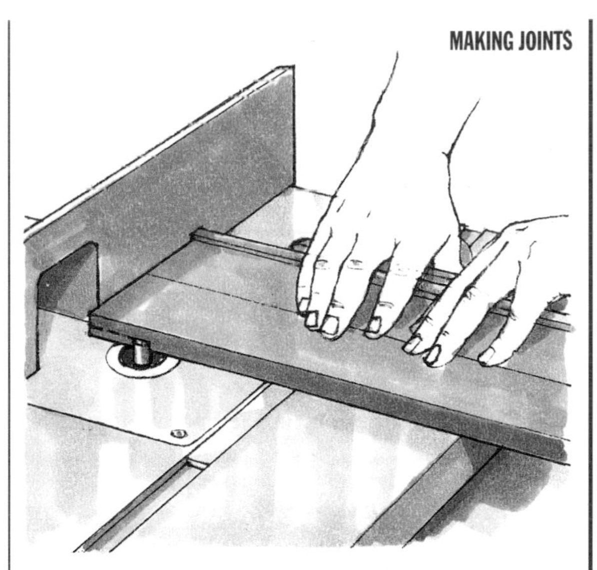

Cutting halvings with an inverted router
Cut corner halving joints on an inverted-router table, using the sliding mitre fence to support the work (see page 71). Butt the end of the workpiece against the table fence to align the shoulder with the cutter, and sandwich a waste batten between the mitre fence and the work to prevent break out. Holding the work securely, feed it into the cutter to machine the shoulder, then slide the wood away from the table fence between passes to remove the waste in stages.

CUTTING IDENTICAL COMPONENTS

Since many projects include a number of identical joints, it pays to plan out in advance which pieces can be cut simultaneously, using slightly larger jigs and a home-made wedge cramp.

Lay the components side by side, and hold them together with the wedge cramp, then clamp the L-shape jig or jigs on the work in the required position to secure the whole assembly to the bench.

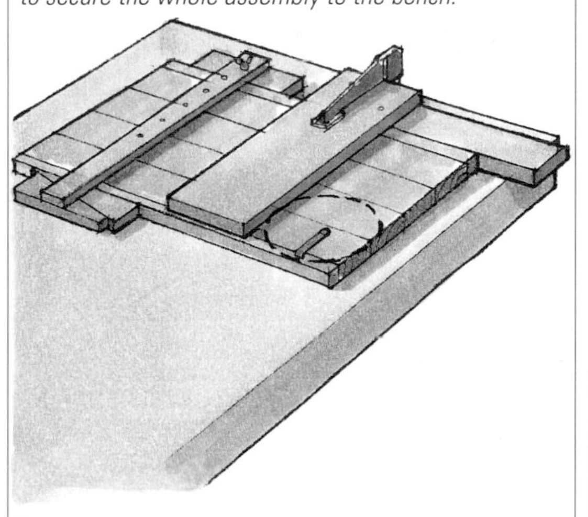

CUTTING HALVINGS WITH A TEMPLATE JIG

For faster, semi-batch production, make a template jig to help you machine halving joints, using a router fitted with a guide bush (see pages 50–1). Revise the recommended dimensions given for the jig to suit your own requirements.

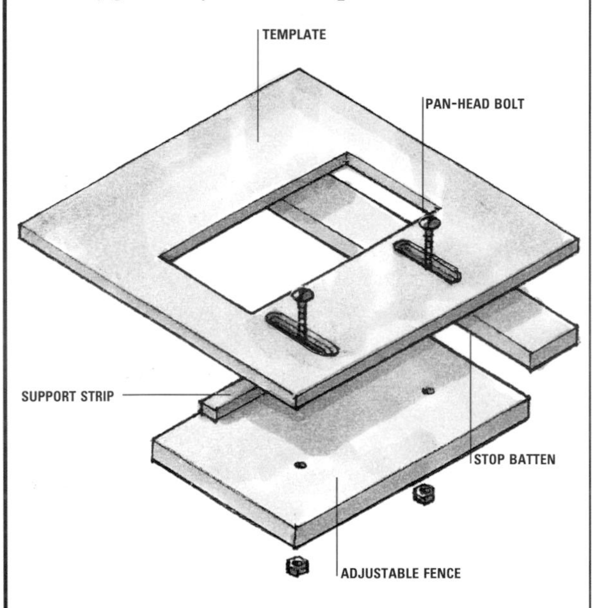

Making a template jig

Cut a 200 x 100mm (8 x 4in) rectangle in the centre of a 300 x 300 x 6mm (12 x 12 x ¼in) MDF template. Cut two 3mm (⅛in) slots at right angles to the edge of the rectangle across the right-hand face. Recess these to a depth of 3mm (⅛in) by 8mm (5⁄16in) wide, to take two 25mm (1in) pan-head bolts. Cut a 300 x 50 x 12mm (12 x 2 x ½in) stop batten, and glue it to the underside of the template, so that it lies flush with the edge of the cutout rectangle.

Cut a 250 x 150 x 12mm (10 x 6 x ½in) adjustable fence. Drill two 3mm (⅛in) bolt holes through the fence, and recess the holes on the underside to allow two hexagonal nuts to be made captive in them. Cut and glue a 200 x 12 x 6mm (8 x ½ x ¼in) MDF support strip to the leading edge of the fence. Mount the fence to the template, using a pair of bolts.

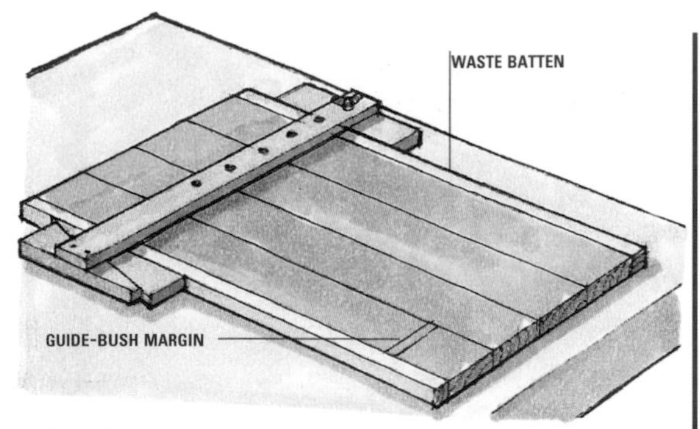

1 Marking out halvings

Mark out the shoulder line on one workpiece, and draw a second line that represents the guide-bush margin (see page 50) behind the shoulder line. Align all components on the bench, backing them up with a waste batten of the same thickness to prevent break out. Clamp the components side-to-side, using a wedge cramp (see also page 37).

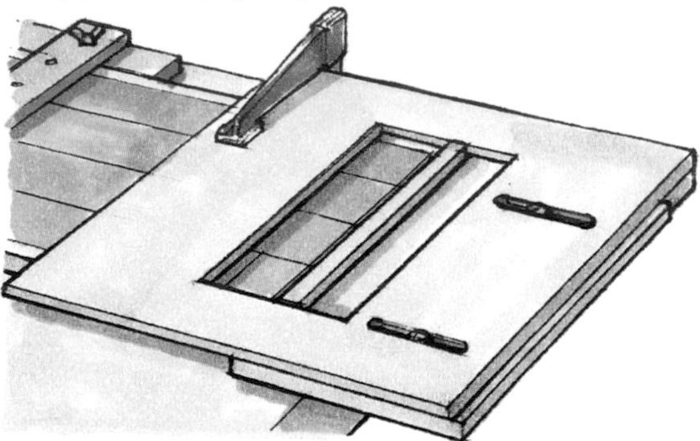

2 Positioning the template jig

Lay the template on the work, butting the stop batten against the waste break-out strip, and align the inside edge of the cutout with the guide-bush-margin line marked on the work. Slide the fence up to within 12mm (½in) of the workpiece ends, then tighten its bolts. Clamp both the jig and work to the bench.

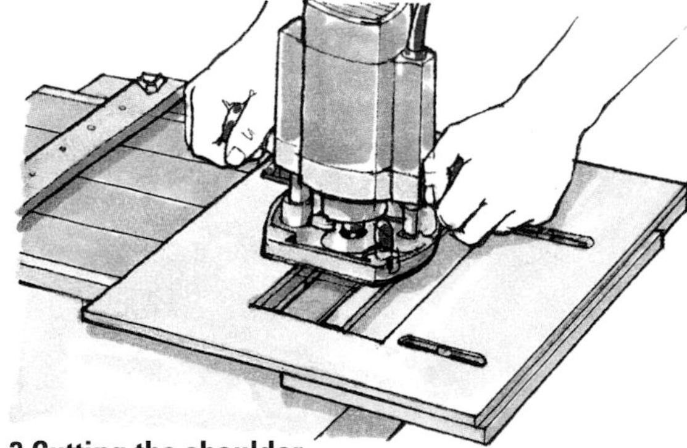

3 Cutting the shoulder

Fit the cutter, set the maximum cutting depth and adjust the turret stop to cut in two or more passes. Machine across the shoulder, running the guide bush against the inside of the cutout, then remove the waste, using the router freehand.

MORTISE-AND-TENON JOINTS

The mortise and tenon is a versatile, exceptionally strong joint, used to construct traditional frame-and-panel cabinets, chairs and tables. In its basic form, the tenon, a tongue cut on the end of a rail, fits in a slot or mortise machined in the leg or stile. All mortises cut with a router have rounded ends. You can square the ends with a mortise chisel, or round over the edges of the matching tenon, using a chisel or rasp.

Cutters

Some woodworkers use straight, single-flute cutters for mortising, but you will achieve better results using an up-cutting spiral cutter to machine narrow mortises, or a step-tooth pocket, or mortise, cutter for slots greater than 12mm (½in) wide (see page 20). Use a long-shank pocket cutter for machining deep narrow mortises; both these types of cutter have good bottom-cutting and waste-clearing characteristics.

Using twin side fences

To machine a mortise in a chair or table leg, for example, fit a pair of side fences to the router to keep the cutter on line (see page 46).

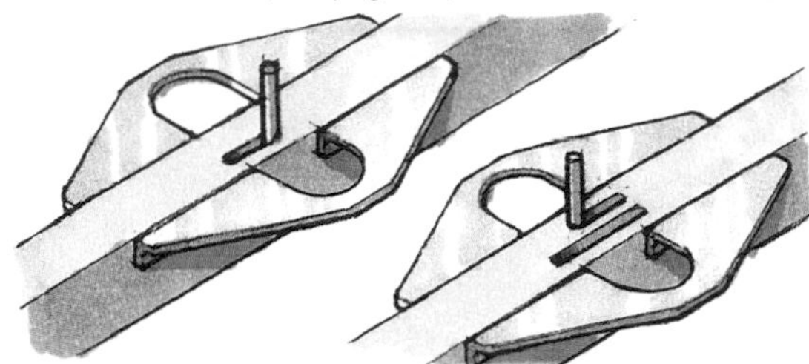

Cutting twin mortises

Use a similar arrangement of fences to cut a pair of parallel twin mortises. Having machined one mortise to the full depth, turn the router around and cut the second mortise in the opposite direction.

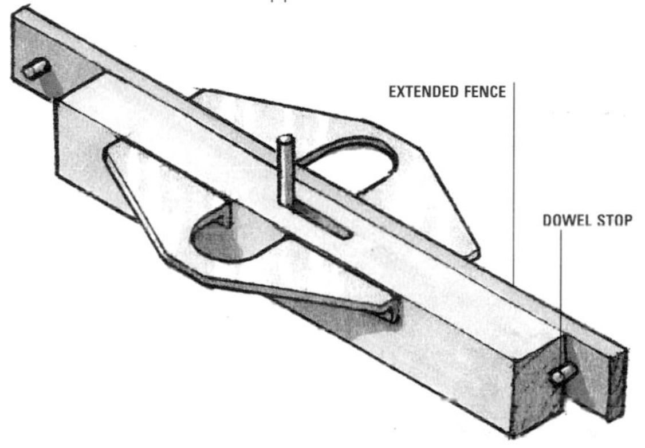

Machining mortises to length

To cut a mortise to length, plunge and raise the cutter according to lines marked on the wood, or clamp stop blocks to the side of the work or bench.

Alternatively, when machining short workpieces, push dowels through holes bored in extended fence faces, to limit the start and finish of each pass.

Cutting mortises for haunched tenons

A short tongue or haunch, which supports the top edge of the rail when a mortise is cut near the end of a leg or stile, fits into a shallow groove cut just above the mortise. For the first few passes, machine a haunch groove along with the mortise, then reset the stops and cut the mortise to its full depth.

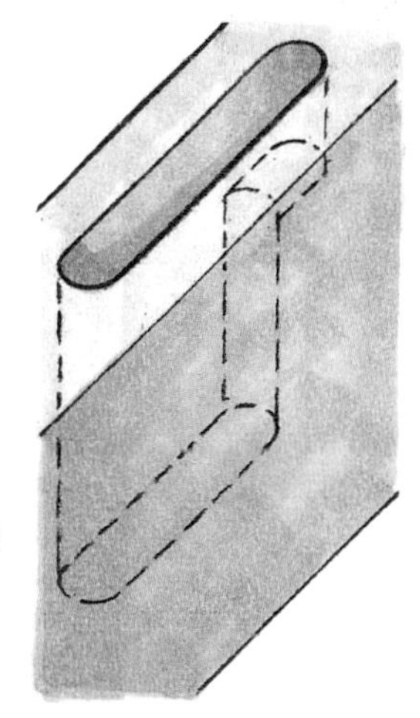

MORTISE TEMPLATES

Instead of using side fences and stops to guarantee accurate mortising, you can cut single and twin mortises using a slot template mounted on the work to guide a router fitted with the appropriate guide bush. You can construct any number of simple templates for specific jobs, or make a multi-purpose template, incorporating adjustable stops and locating battens.

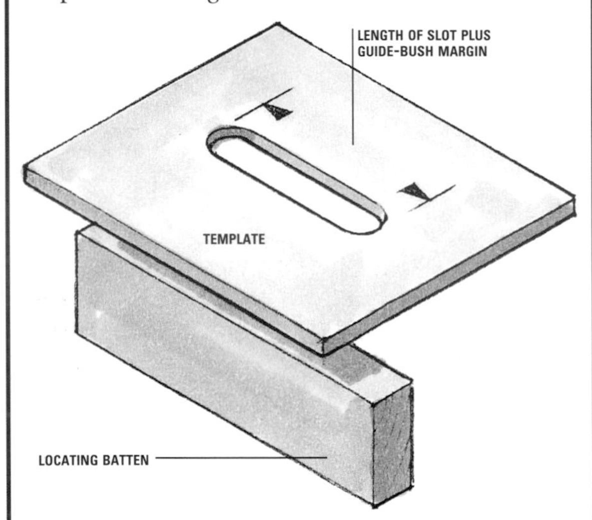

Making a simple mortise template

Cut a 6mm (¼in) MDF template, about 150mm (6in) long and as wide as the router base plate. Using a cutter that is the same diameter as your chosen guide bush, cut a slot down the centre of the template. This slot should be as long as the mortise, plus the guide-bush margin (see page 50) at each end.

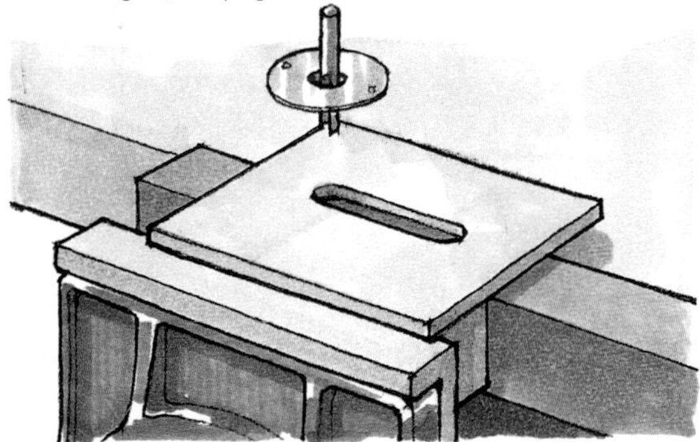

Adding a locating batten

In order to locate the template accurately on the edge of the workpiece, glue a stout batten to the underside to act as a fence which will align the slot with the mortise position. Clamp the template to the work and machine the mortise.

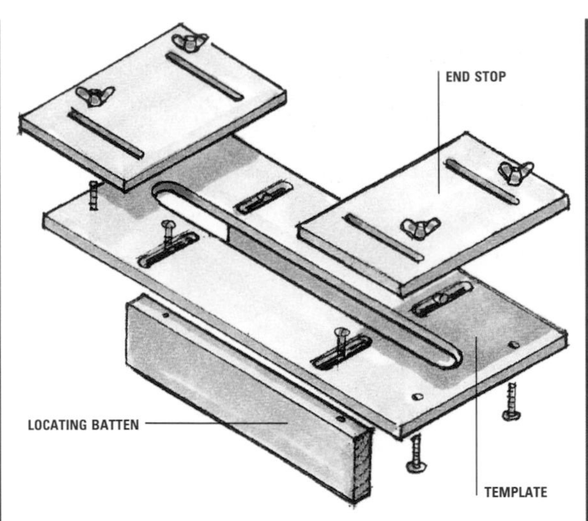

Making an adjustable template

To make a template that can be used for mortising any workpiece, cut the MDF panel as before, but increase the overall length to 300mm (12in) and the length of the slot to within 25mm (1in) of each end. Cut the slot to fit a guide bush that is large enough to accommodate any cutter you will use for mortising. Cut two pairs of recessed slots for attaching locating battens, and drill two recessed holes at each end of the template to take the end-stop screws.

Cut end stops from 6mm (¼in) MDF, and machine two slots for the adjusting screws in each stop. Fit the stops to the template with pan-head bolts, washers and wing nuts.

Cut the locating battens from 38 x 9mm (1½ x ⅜in) timber, and screw them to the template with pan-head screws and washers.

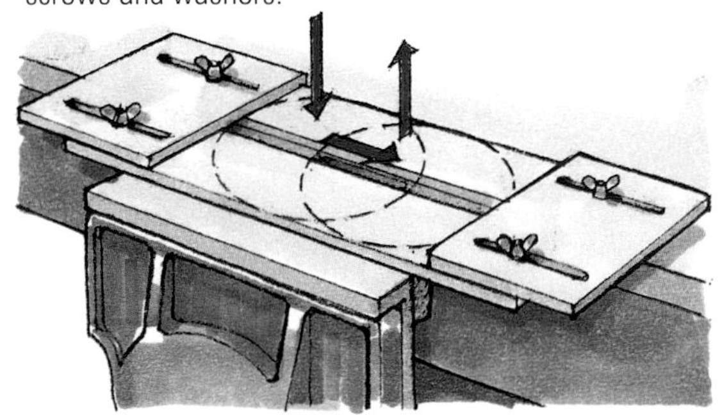

Using an adjustable template

Adjust the locating battens to align the template slot with the mortise position, and clamp the template to the work. Adjust each end stop to restrict the passage of the router according to the length of the mortise. With the base plate butted against the first stop, switch on and plunge the cutter into the wood. Feed the router forward until the base plate comes to rest against the other stop, release the plunge lock to lift the cutter, and switch off.

MORTISING WITH A TABLE-MOUNTED ROUTER

Cutting a mortise on an inverted-router table means lowering the timber onto the cutter (see page 78). When machining small or short workpieces, make up a guard that also incorporates end stops that allow you to cut a mortise precisely to length.

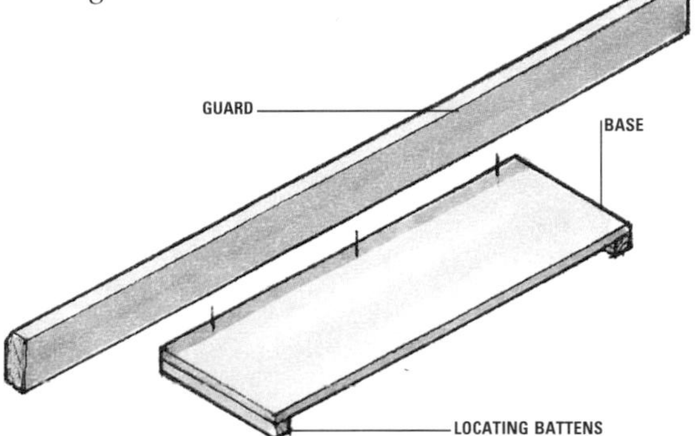

Making a stopped guard

Cut a 150mm (6in) wide base for the guard from 6mm (¼in) MDF, making it the length of the router table plus 25mm (1in). Glue 12 x 12mm (½ x ½in) battens to the underside of the base, flush with each end. These battens locate over the edges of the router table. Cut an 18mm (¾in) thick guard, about twice the length of the table fence. Make the guard at least 12mm (½in) higher than the fully extended cutter, less 6mm (¼in) to allow for the base. Glue and screw the base to the bottom edge of the guard, making a flush, right-angle joint. Plane a small chamfer along the top edge of the guard, on the side nearest the fence.

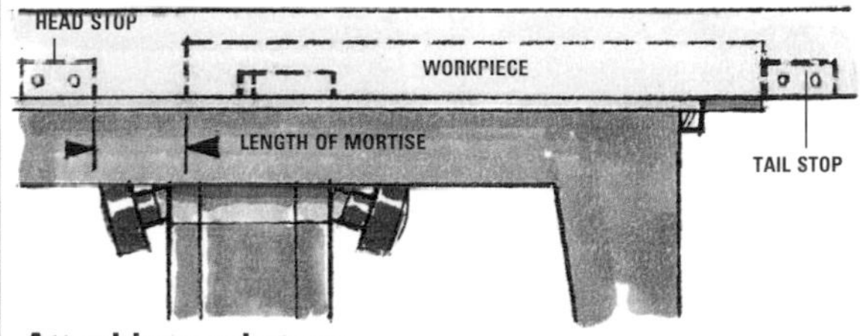

Attaching end stops

Cut two stops from 50 x 12mm (2 x ½in) hardwood. Clamp or, better still, screw the stops to the fence-side of the guard. With the bottom end of the work butted against the tail stop, the cutter should align with the top of the mortise. By the time the top end of the work comes up against the head stop, the cutter should have machined the entire length of the mortise. When cutting a mortise near the top of a leg or stile, leave plenty of waste at the end of the workpiece, to keep your fingers clear of the cutter.

Adjusting the guard

Adjust the table fence to align the cutter with the centre of the mortise. Lay a workpiece against the table fence and slide the guard up to the work, allowing for a snug sliding fit. Secure the guard and set the cutter height for the first pass.

Cutting the mortise

Butt the rear end of the work against the tail stop and, holding the front end of the workpiece with your fingertips only, carefully lower the wood onto the cutter. Feed the work forward until it comes to rest against the head stop. When cutting several pieces, repeat this procedure on each workpiece before re-setting the cutter height for the next pass.

CUTTING TENONS

You can cut tenons using a method similar to that described for making halving joints (see pages 88-90) – cut one face, turn the workpiece over and cut the second one. It is generally best to cut square or sloping haunches by hand after you have machined the tenons.

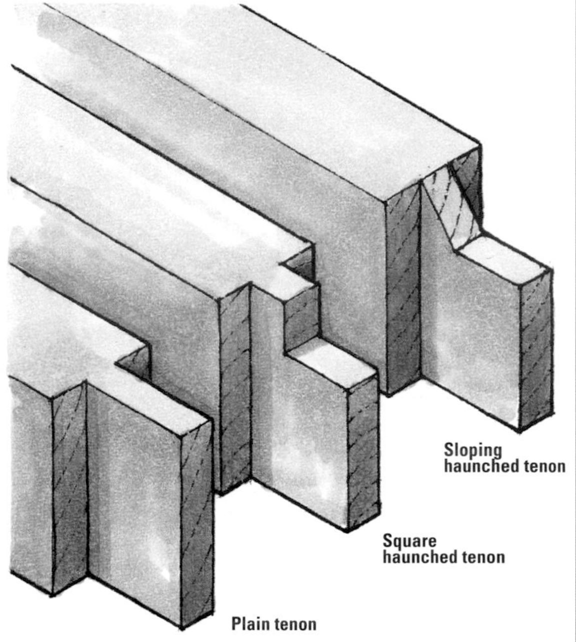

Using a right-angle jig

Make an L-shape jig by gluing and screwing together two pieces of MDF at right angles (see page 88).

Lay the jig on the rail and clamp the assembly to the bench. Set the maximum depth of cut and, if necessary, adjust the turret stop to cut in two or more passes. Cut the shoulder first, then remove the remaining waste by cutting freehand.

When cutting a narrow tenon, it may be necessary to support it on a piece of timber, to prevent the wood vibrating as you cut the second face.

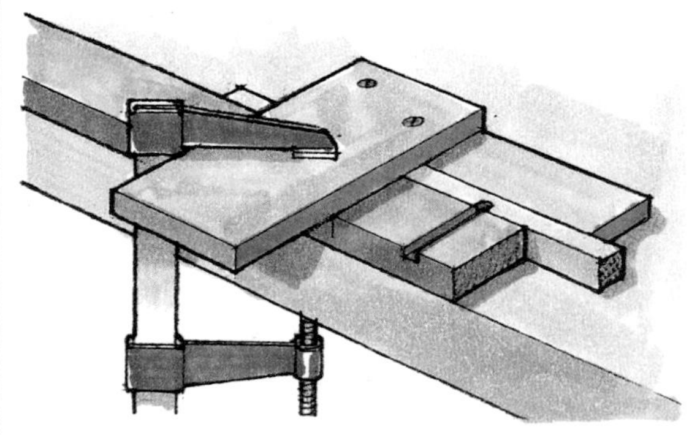

Cutting on a router table

Using the sliding mitre fence to support the work, butt the rail end against the table fence to align the shoulder line with the cutter. Lay a waste batten between the mitre fence and the work to prevent break out on the rear face. Feed the work forward to cut the shoulder, then slide the rail away from the fence to remove the waste in stages.

Some inverted-router tables are fitted with a vertical sliding cramp attached to the fence, for cutting tenons on short rails.

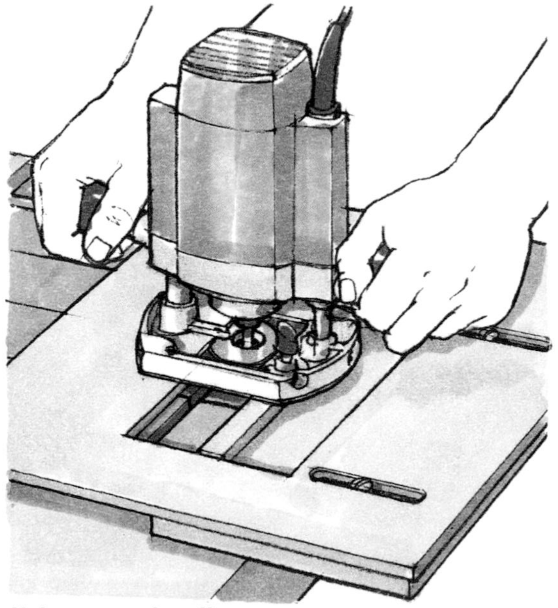

Using a template jig

You can also cut tenons on the adjustable template jig (see page 90). Machine the shoulders on one side of the rails, then turn them over, replace the template jig and cut the other set of shoulders.

DOWEL JOINTS

Though a dowel joint is nothing more than a reinforced butt joint, it rivals the mortise-and-tenon joint for strength. It is also more versatile, being equally suitable for constructing frames from solid wood and carcasses from man-made boards. Because the dowels must be aligned accurately in both halves of the joint, it is best to use a template or a pair of side fences to position the router on the work. Use special-purpose router dowel drills, rather than straight cutters or ordinary power-drill dowel bits.

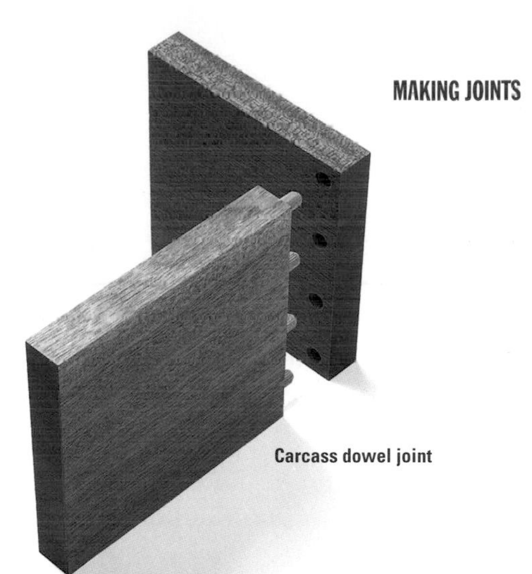

Carcass dowel joint

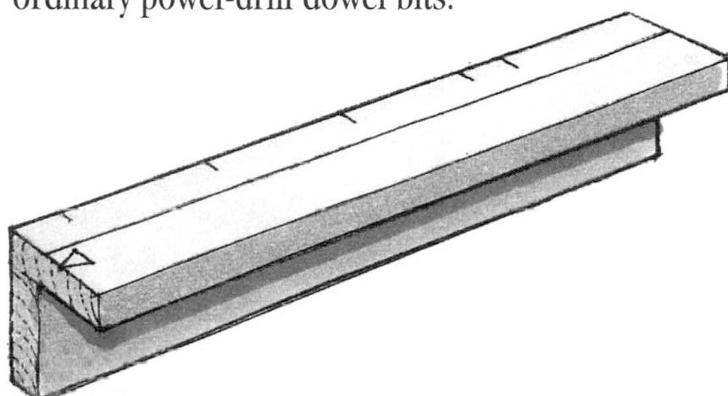

Making a dowel-spacing batten

A simple right-angle guide will help you position a dowel drill quickly and accurately along the edge of a workpiece or across a panel. Cut two lengths of 75 x 18mm (3 x ¾in) wood, slightly longer than the width of the workpiece. Mark out the width of the work and required dowel positions on the top face of the guide batten. Some woodworkers apply and mark a strip of masking tape, which can be replaced from time to time as work dictates.

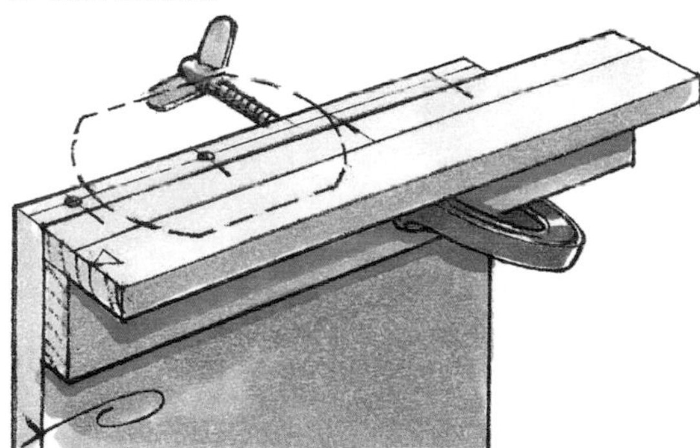

Edge-dowelling with a spacing batten

Mark out the centre line along one edge of the workpiece. Set the work upright in a vice and clamp a dowel-spacing batten flush with the end and face edge. Using a try square, transfer the dowel centres onto the centre line, then set the router side fence to align the dowel drill. Bore each hole in turn.

Boring dowel holes across the face of a board

Mark the centre line of the joint. Position the dowel-drill point on the line and mark the edge of the router base plate. From this mark, draw a line square across the board. Place a spacing batten against the squared line, with its end flush with the board's face edge.

Mark the cutter centre on the edge of the router base (many routers are supplied already marked). Aligning this mark with the dowel centres pencilled on the spacing batten, plunge-cut each hole in turn.

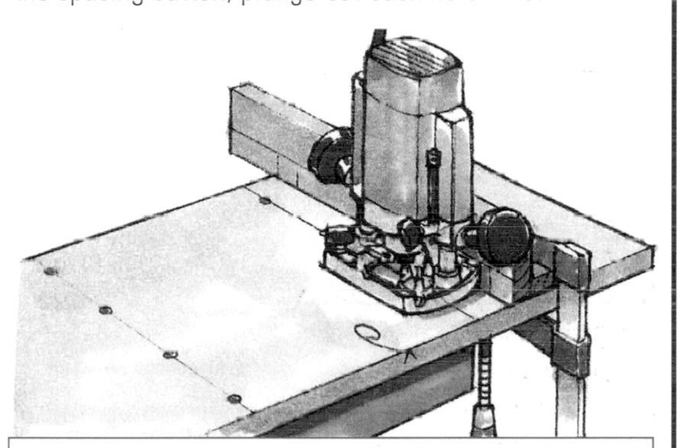

USING TWIN SIDE FENCES

Adding a second side fence ensures that a dowel can be positioned accurately on the edge of a workpiece. Clamp both halves of the joint in a vice, face sides out and ends flush. Using a marking gauge, score a centre line down each board, then measure off the positions of the dowels.

With the spur of the dowel drill on one centre line, slide both fences up to the work, leaving minimum side play (see page 46). Set the depth gauge to allow the drill to be plunged to the full dowel depth – half the length of the dowel plus 1.5mm (1⁄16in).

Work along one edge, positioning and plunge-boring each hole in turn. Turn the router around and repeat the operation along the other board.

MAKING A DOWELLING TEMPLATE

Dowelling with a simple jig is faster and slightly more accurate than using a dowel-spacing batten. Dowelling templates can be cut from 6mm (¼in) MDF or, better still, from transparent, shatterproof plastic. Cut the template to suit the size of the work, allowing a wide margin to provide adequate support for the router, and for fitting adjustable locating battens.

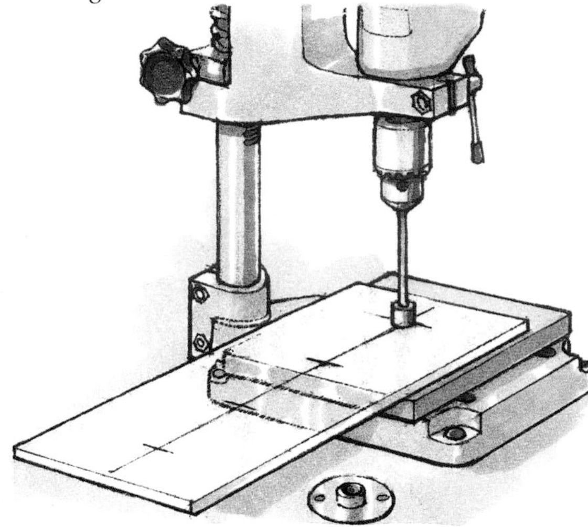

1 Marking and boring dowel centres
Set out the dowel spacing along a centre line drawn or scribed across the template – score the line on the underside of a plastic template. Drill each dowel-spacing hole to match the diameter of the required guide bush, which should allow a minimum 1.5mm (1⁄16in) clearance around the dowel drill.

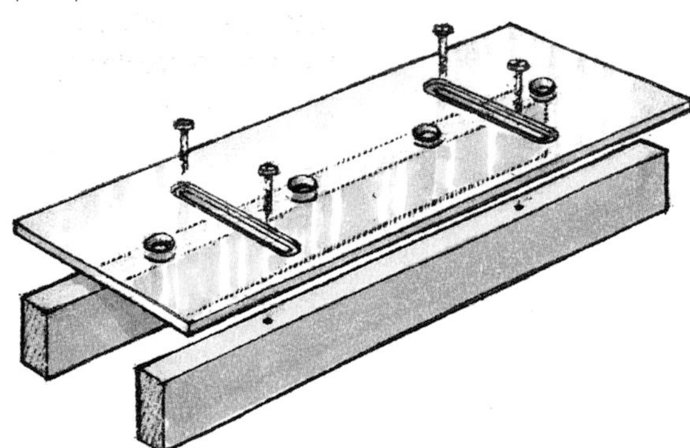

2 Fitting locating battens
Mark out and cut two 3mm (⅛in) slots across the template. Recess or countersink each slot on the top face. Cut two 50 x 18mm (2 x ¾in) battens and fit them beneath the template, using pan-head or countersunk screws that pass through the recessed slots.

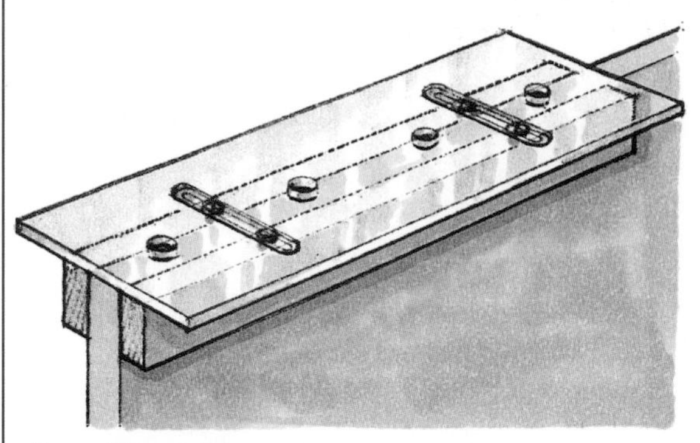

Dowelling edges
Align the dowel-spacing holes with a centre line scored down the edge of the workpiece. Slide both locating battens up to the work and tighten the screws. Fit the guide bush in each hole in the template in turn, and plunge the router.

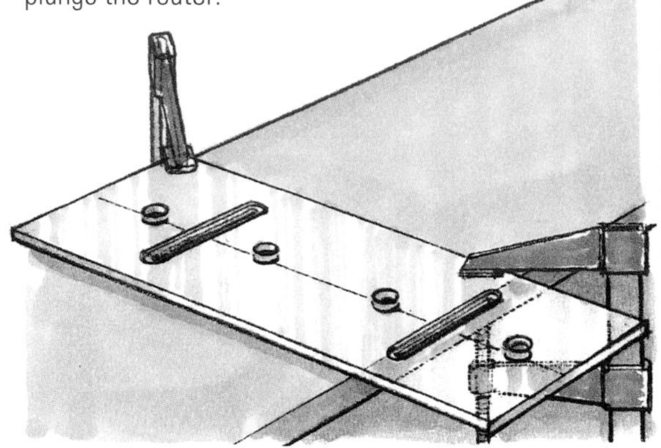

Dowelling across the face
Mark out a centre line across the face of the workpiece. Remove the locating battens from the template and, having aligned the dowel-spacing holes with the centre line, clamp the template to the work. Bore each dowel hole in turn.

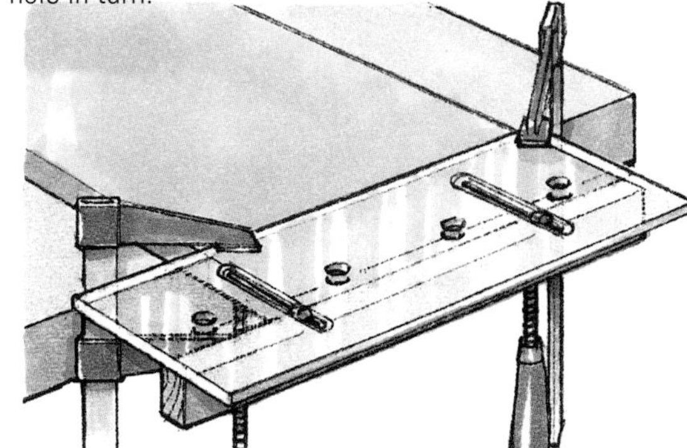

Making corner joints
To dowel corner joints, first bore the holes in the end of one panel as described above, then remove one locating batten and butt the remaining batten against the end of the other component, securing the template with a cramp.

DOWELLING JIGS

On some factory-made dovetail jigs, the template can be replaced with a metal plate that is perforated for boring dowel holes. The jig will accommodate boards up to 610mm (2ft ⅜in) wide, and between 12 and 30mm (½ and 1⅛in) thick. It is also possible to edge-joint longer workpieces by clamping them to the face of the bench and moving the jig along the boards.

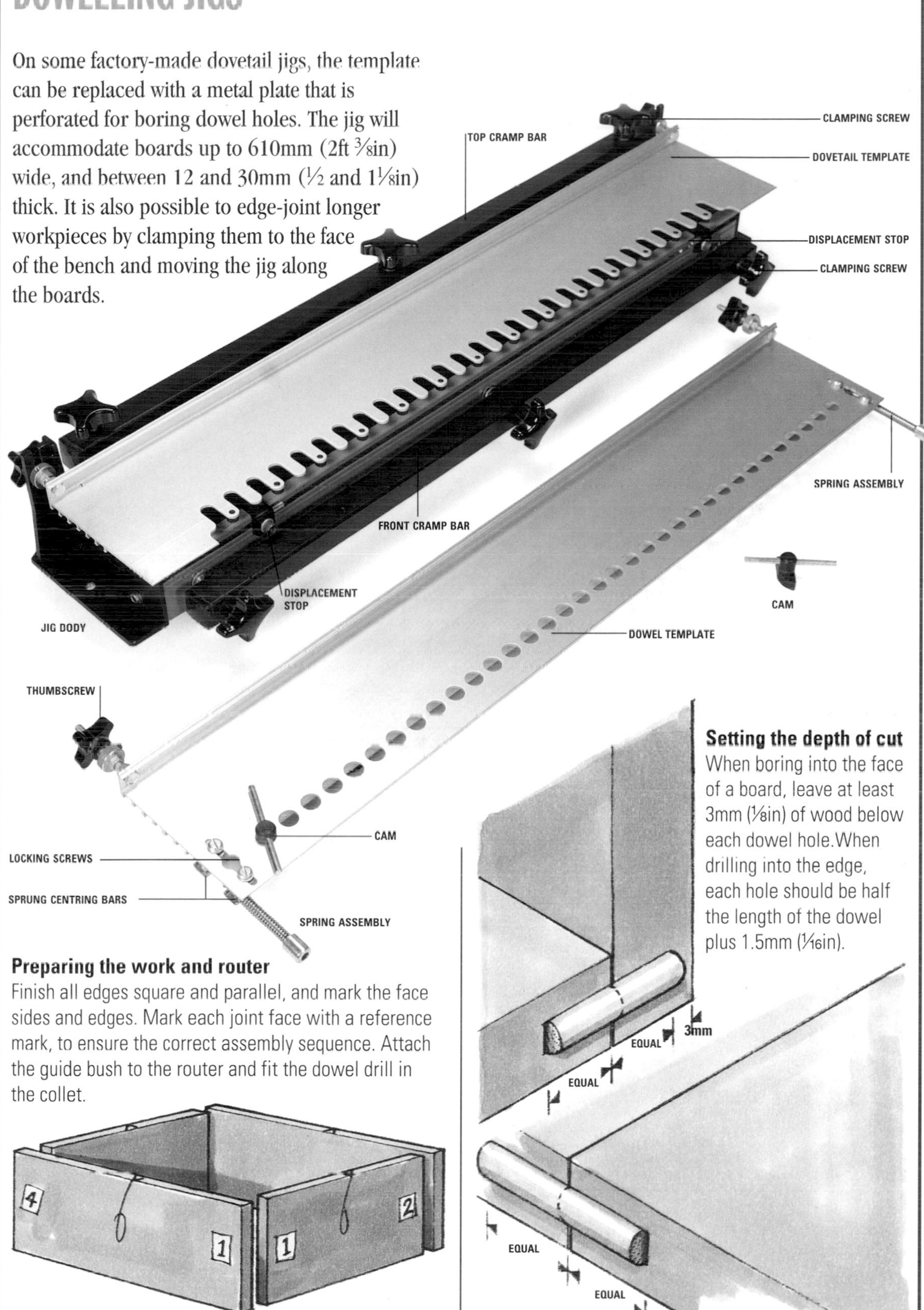

Setting the depth of cut

When boring into the face of a board, leave at least 3mm (⅛in) of wood below each dowel hole. When drilling into the edge, each hole should be half the length of the dowel plus 1.5mm (1⁄16in).

Preparing the work and router

Finish all edges square and parallel, and mark the face sides and edges. Mark each joint face with a reference mark, to ensure the correct assembly sequence. Attach the guide bush to the router and fit the dowel drill in the collet.

Dowelling panel ends

Whether you intend to use a jig to make a corner joint or a T-joint, for a fixed partition or shelf, you need to bore dowel holes in the end of one of the components.

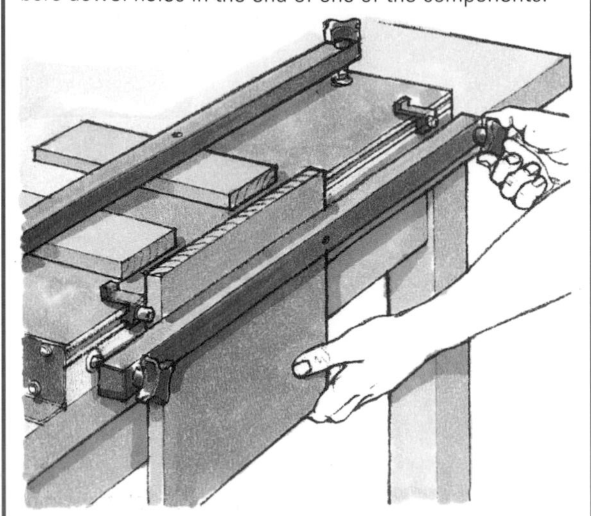

1 Clamping the work in the jig

Place two pieces of waste timber, the same thickness as the workpiece, on the top face of the jig. Slip the workpiece behind the front bar cramp, temporarily set it 6mm (¼in) above the waste timber, and tighten the bar cramp. To accommodate a wide board, remove the centre clamping screw entirely.

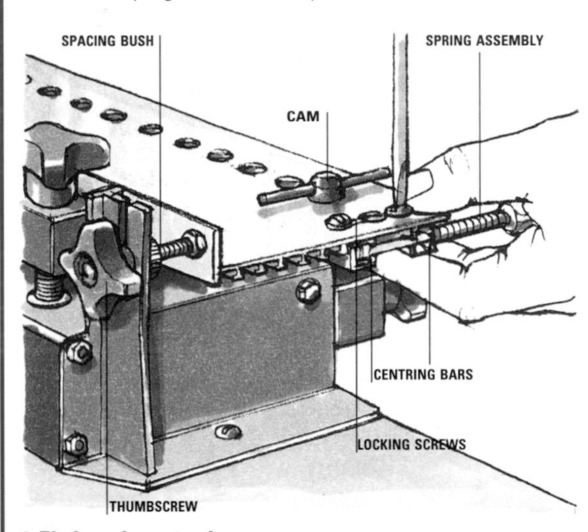

2 Fitting the template

When fitting the template over the end of the workpiece, cam-operated sprung centring bars grip the board, automatically centring the row of dowel holes. Lower the template to rest on the face of the waste timber. Secure the centring bars by tightening their locking screws so that you can then remove their spring assemblies. Secure the template by adjusting the spacing bushes and tightening the thumbscrews at the back of the jig.

3 Centring the row of dowel holes

Slacken off the front bar cramp and move the workpiece sideways to position dowel holes an equal distance from both edges of the board, with sufficient wood around each dowel for a strong joint. Re-tighten the cramp, and slide the left-hand displacement stop against the board edge. Tighten the locking screw.

4 Lowering the workpiece

Lift off the template, without disturbing the settings, and remove the cams. Set the workpiece flush with the waste pieces, check the edge is against the stop, then replace the template.

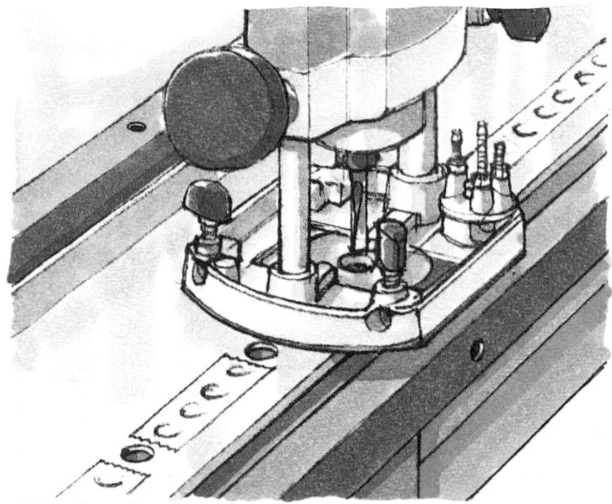

5 Boring dowel holes

Mask off with tape any template holes that you do not require, and locate the guide bush in the others to plunge-cut a row of dowel holes. Before boring the opposite end of the workpiece, use the drilled end to set up the right-hand displacement stop.

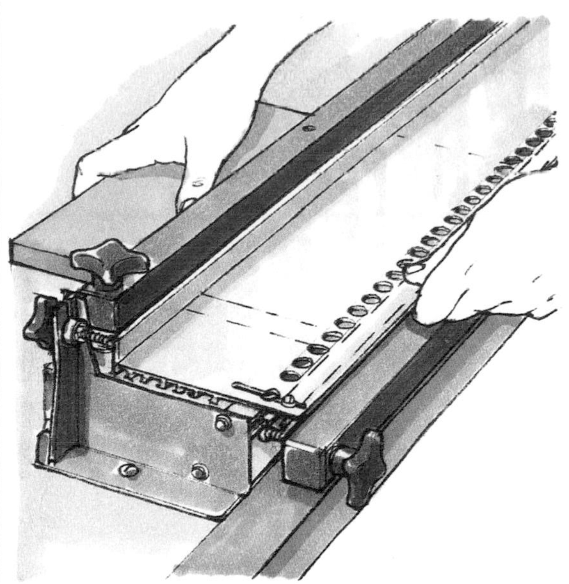

Making a corner joint

After dowelling the end of one component, remove the rear centring bar from beneath the template, but leave the front bar in place. Position the workpiece under the top bar cramp and replace the template. Slide the workpiece forward, up to the remaining centring bar, and adjust the work sideways until it butts against the displacement stop. Secure the template and tighten the bar cramps.

Set the depth of cut and then plunge-bore each hole in turn.

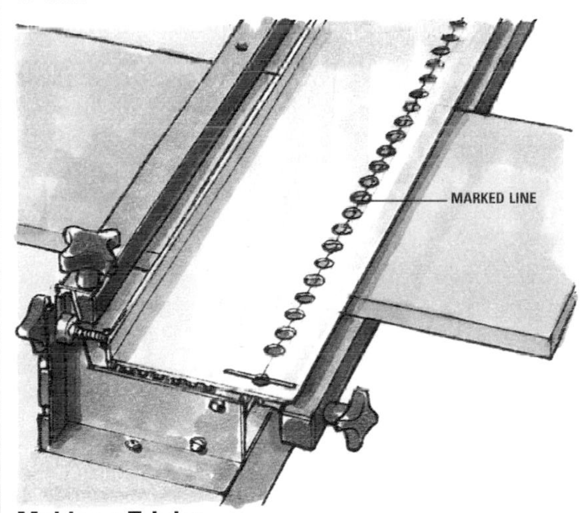

Making a T-joint

Dowel the end of one workpiece (see opposite), then remove both centring bars from the template. Draw the centre line of the joint across the face of the board. Slide the board under the top bar cramp and template, and align the centre line with the row of holes in the template. Tighten all cramps and bore the dowel holes.

DOWELLING LONG EDGES

Use a small waste piece of timber the same thickness as the workpiece to set up the template, centring the row of dowel holes (see opposite). Remove the displacement stops, front bar cramp and template from the dowelling jig.

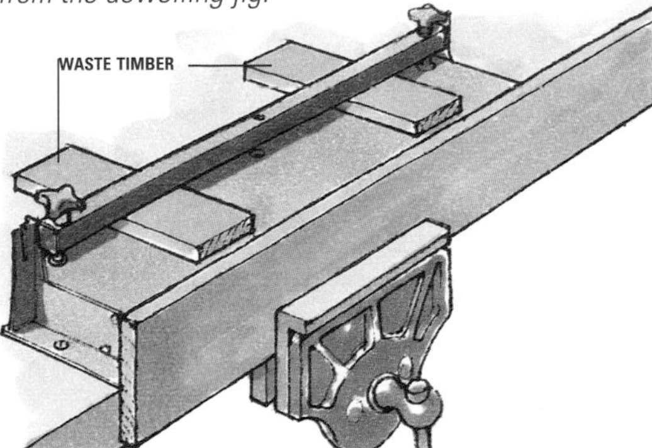

1 Setting up the work

Stand the jig on the bench. Fit two pieces of waste timber, the same thickness as the work, beneath the top clamping bar, and clamp the workpiece to the front of the bench. Align its top edge flush with both waste pieces clamped in the jig.

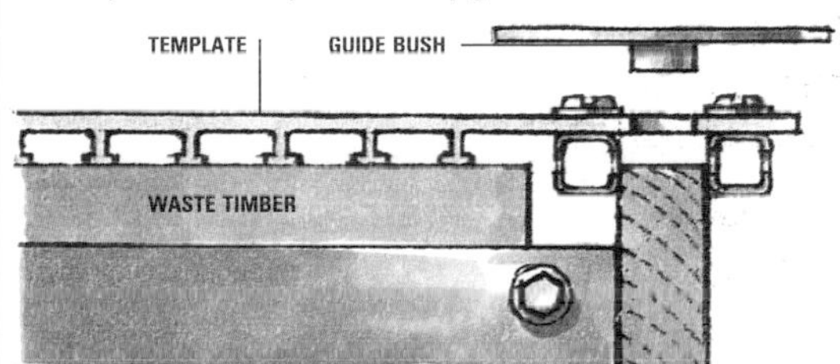

2 Fitting the template

Fit the template over the edge of the workpiece, locating the centring bars on both sides of the board.

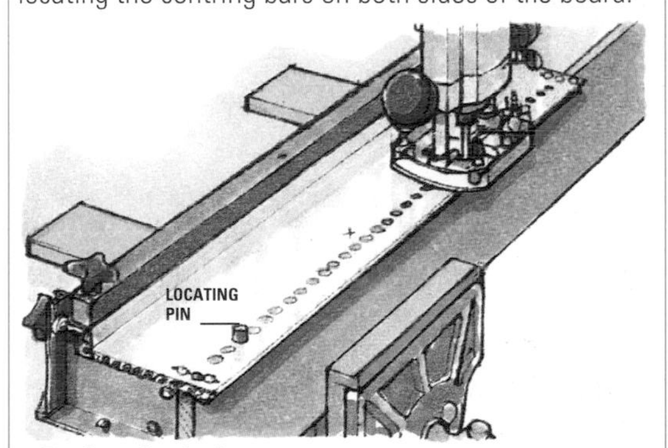

3 Boring the holes

Bore the first hole and insert a metal locating pin, then plunge the router through the remaining holes in the template. Pull out the locating pin, slide the jig sideways and insert the pin through the template into the last hole you bored. Continue boring holes until the entire row is complete.

BISCUIT JOINTS

A different form of reinforcement, employing flat wooden discs or 'biscuits' instead of dowel rods, creates extremely strong butt joints. The slots to accommodate the biscuits are usually cut with a special-purpose miniature circular saw (biscuit jointer), but you can machine similar joints just as effectively in solid timber and man-made boards, using a power router.

Corner-butt joint

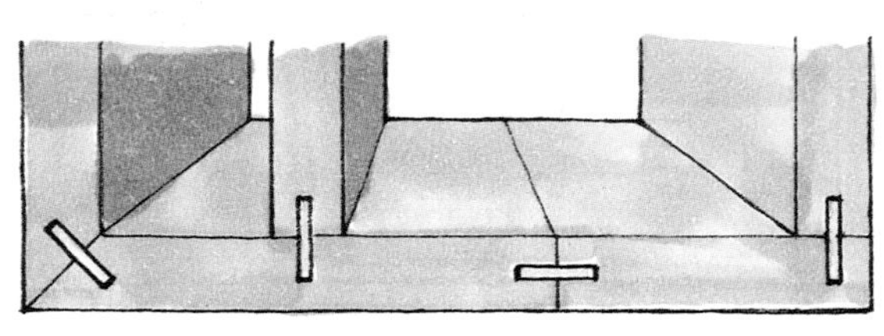

Types of biscuit joints

You can make corner- and T-butt joints, mitred and edge-to-edge joints, in both solid wood and man-made boards. Biscuit joints are mostly used for framing or in cabinetmaking.

Biscuit dowels

Biscuit dowels, made from compressed hardwood chips, are glued into semicircular recesses cut into both mating faces of the joint. When assembled, the glue causes the compressed fibres to swell within the recesses, making a tight fit between each biscuit and the workpiece. This swelling is unlikely to split the wood or board unless either is particularly thin.

Biscuits are commonly available in three standard sizes: No. 0, to suit boards from 9 to 12mm (3/8 to 1/2in) thick; No. 10, for boards 13 to 18mm (9/16 to 3/4in) thick; and No. 20, for boards over 19mm (13/16in) thick. All biscuits are 4mm (5/32in) thick.

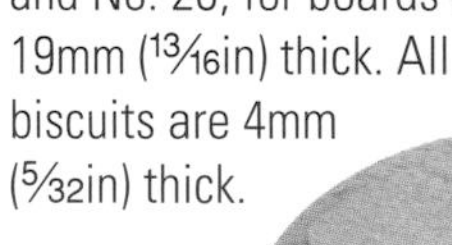

Hardwood biscuits

BISCUIT-JOINT CUTTERS

Biscuit-dowel recesses are cut along the edge or end of a board, using a 4mm (5/32in) slotting cutter mounted on a threaded arbor. The arbor is fitted with one of three different guide bearings, 16, 18 and 22mm (5/8, 3/4 and 7/8in), depending on the depth of cut required for a particular size of biscuit. Always fit the guide bearing before fitting the arbor in the collet.

Cutters for slotting across the face

Since a grooving cutter cannot be used to machine across the face of a board, you will need to utilize a 4mm (5/32in) diameter single- or two-flute straight cutter in order to recess one half of a T-joint or a mitre joint.

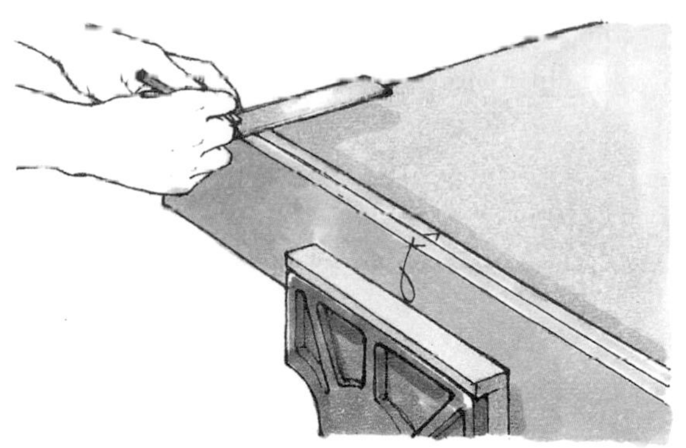

1 Marking out the work for edge-jointing
Clamp the two workpieces together, face sides out and joining edges flush. Mark the centres of the biscuit-dowel recesses across both boards.

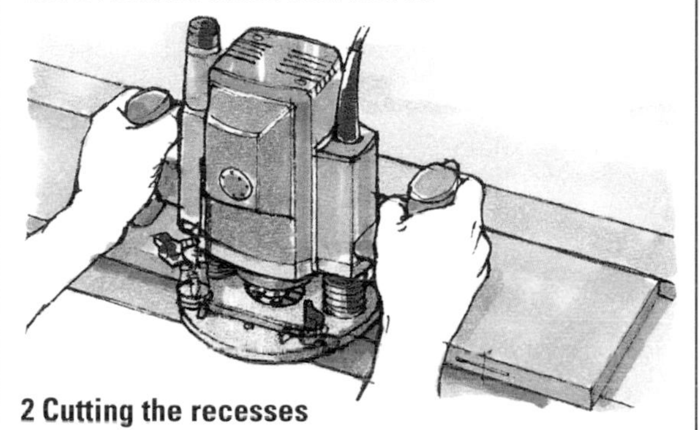

2 Cutting the recesses
Clamp one board overhanging the edge of the bench, face-side up. Rest the base plate on the face of the work, set the depth stop and lock the router with the slotting cutter centred on the edge of the workpiece.

Centre the cutter on the first mark and feed the router straight into the edge of the board until the bearing prevents it cutting deeper. Repeat for each biscuit recess.

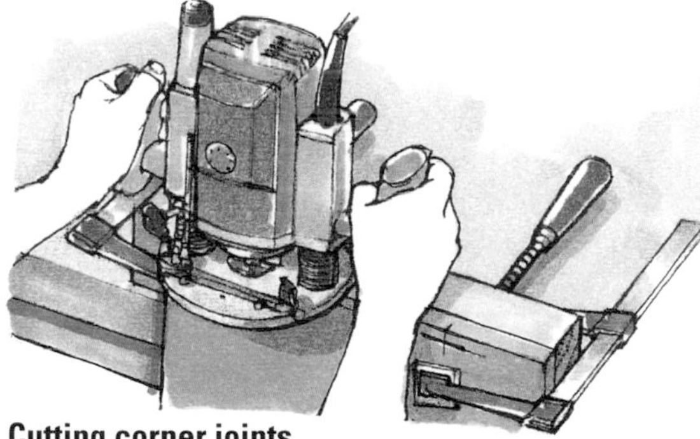

Cutting corner joints
Having recessed the end of one component, set the other half of a corner joint upright in a bench vice.

Clamp a batten flush with the end of the board to keep the router base plate level. Cut each recess as described for edge-jointing (see above). If the board is too long to stand upright in the vice, clamp it face-up on the bench and cut the recesses with the router body held horizontally.

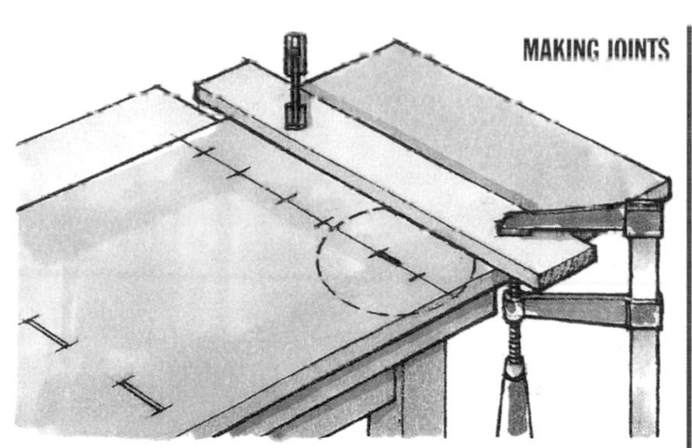

Cutting T-joints
To insert biscuit dowels across the face of a panel, fit a straight cutter in the router and set the depth slot to cut each recess 1.5mm (1/16in) deeper than required for the actual biscuit.

Draw the centre line of the joint across the work, and mark the biscuit recesses along it. Clamp a straightedge guide or T-square (see pages 47–8) across the work to guide the router, and cut each slot in turn.

CUTTING MITRE JOINTS

Insert biscuits in a mitre joint using a straight cutter. Make a slot-cutting template, fitted with a 100 x 50mm (4 x 2in) adjustable clamping batten. Plane one edge of the batten to 45 degrees. Cut a 150mm (6in) wide template from 6mm (¼ in) MDF, and machine a slot along the template to accommodate a 10mm (3/8in) diameter guide bush. Cut two recessed slots across the template to take pan-head screws for attaching the template to the batten.

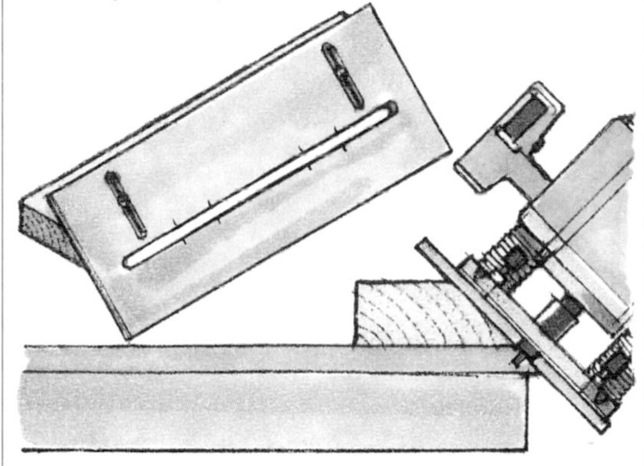

Cutting the slots
Clamp the batten to the face of the workpiece, with the template against the mitre face. Adjust the template to align the edge of the slot with the heel of the mitre. Mark the length of each biscuit slot on the face of the template so that they can be set off against the router base plate. Plunge-cut each biscuit slot in turn.

TONGUE-AND-GROOVE JOINTS

Tongue-and-groove joints are mainly used for constructing wide solid-wood panels by joining boards edge-to-edge. However, you can adapt the joint to make right-angle joints. Joints are made either with an integral tongue to fit in a groove along the mating edge, or as two grooved edges to accommodate a loose tongue.

Edge-cutting operations are generally carried out using a self-guiding slotting cutter or the appropriate straight cutter in conjunction with the router side fence.

Clamping for edge-jointing

Clamp the workpiece on edge in the vice or, using G-cramps, against the edge of the bench. Arrange suitable support at the ends and midway on long workpieces, to prevent bowing. Ensure that any cramps will not interfere with the router and that the side fence has adequate clearance above the vice.

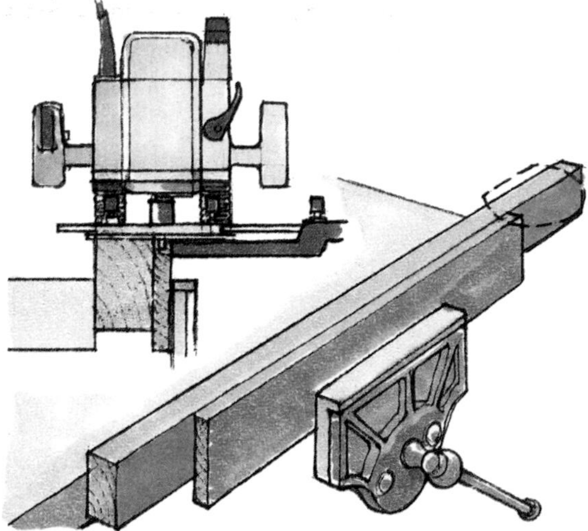

Clamping narrow work

To provide additional support, clamp a thick squared batten alongside a thin workpiece, with the top edges flush, to help keep the router level. A batten longer than the workpiece allows the router to be fed smoothly onto and off the jointing edge. Alternatively, fit a pair of side fences to steady the router and keep the cutter on line (see page 46).

Cutting edge grooves using slotting cutters

Arbor-mounted slotting cutters can be used to machine edge grooves (see page 59). Set the depth of cut by mounting the appropriate guide bearing above or below the grooving cutter.

Most edge grooves can be cut in one pass using these cutters; with hard or troublesome timbers, adjust the side fence or use a larger-diameter guide bearing to limit the depth of the initial pass.

GROOVING ON A ROUTER TABLE

Cutting grooves on an inverted-router table (see page 78) is generally easier than using a hand-held router. Fit the appropriate-diameter cutter in the collet, and set the height above the table to make the first pass. Adjust the table fence to centre the cutter on the edge of the board. If possible, fit a horizontal pressure cramp (see page 74) against the face of the timber. Cut the groove in several passes to the full depth.

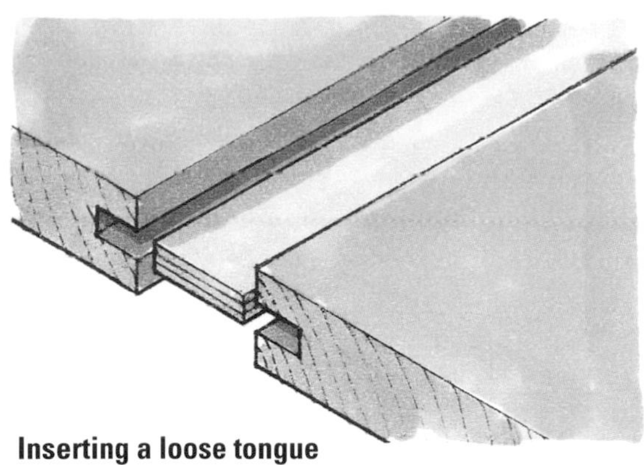

Inserting a loose tongue
To join two grooved workpieces, insert a glued plywood tongue. Alternatively, cut one from the edge of a solid-wood board, using a circular-saw table.

Forming an integral tongue
To cut an integral tongue that will match a grooved workpiece, first machine a rabbet on one side of the board, then turn it over and machine a second, identical rabbet, leaving a tongue centred on the edge of the board. Either use a table-mounted router (see page 76) or a self-guiding rabbet cutter in a hand-held router (see page 61).

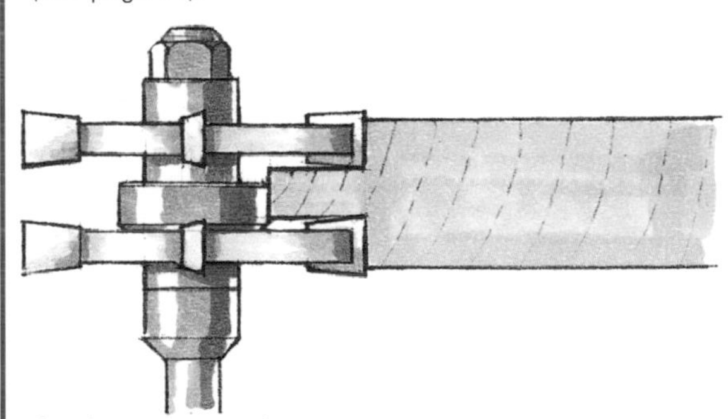

Cutting a tongue in one pass
To halve the time it takes to cut an integral tongue, use a pair of slotting cutters mounted on a single arbor (see page 25). Separate the cutters with a suitable-size guide bearing that determines the length and thickness of the tongue. When machining very hard or trouble-some timbers, use the side fence or a larger-diameter guide bearing to limit the depth of the initial pass.

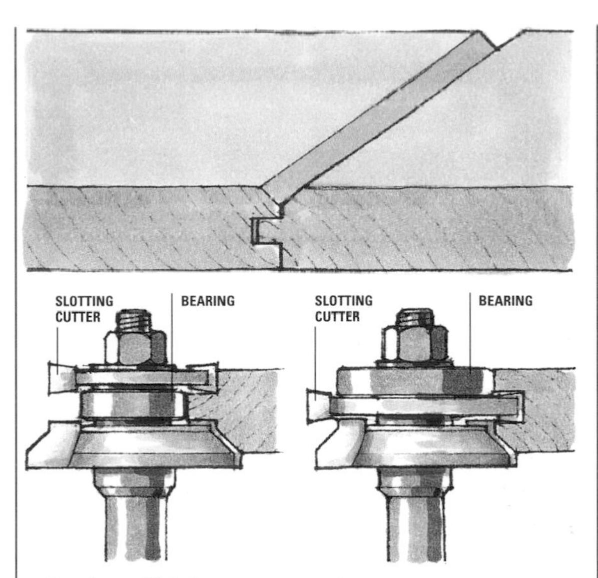

Cutting a V-joint tongue and groove
'Match-lining' cutter sets are available for machining V-joint tongue-and-groove edges. Each set consists of three cutters – two slotting cutters of different diameters, and a cutter ground to a stepped-chamfer profile. Two different-diameter guide bearings are used to vary the depth of cut.

Assemble the cutters to cut one side of the joint profile, and then swap the slotting cutters and bearings to machine the matching reverse profile.

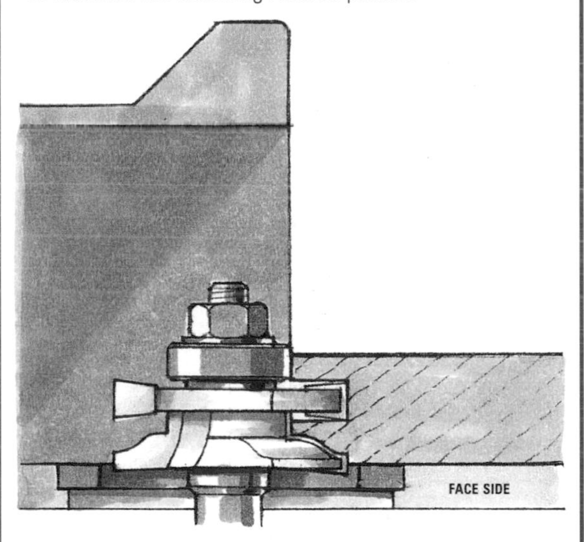

Cutting a decorative-edge tongue and groove
Use profile-scribing cutters (see page 26) to shape one or both face edges of a tongue-and-groove joint. Run the workpiece against the table fence of an inverted-router table.

You can produce similar joints with a hand-held router by machining the tongue-and-groove joint and cutting the appropriate moulding afterwards.

BEAD AND COVE JOINTS

Similar in principle to the standard tongue-and-groove joint, bead and cove joints are special-purpose edge-to-edge joints. Whenever the timber size and profile allow, fit an extended sub-base to the router to help keep it level throughout the operation (see page 30).

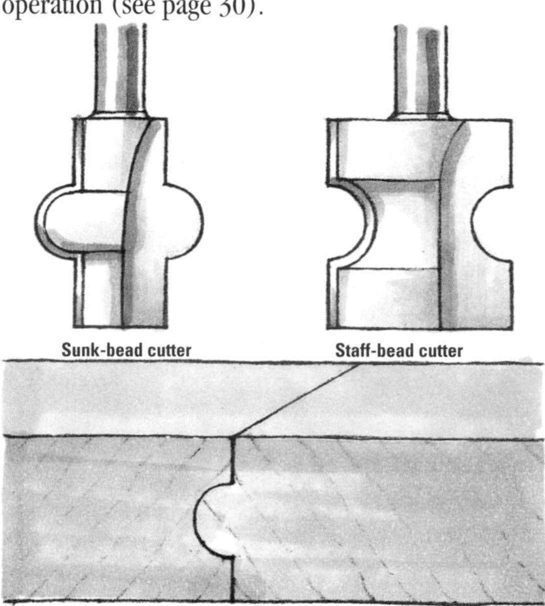

Bead joints
Bead joints, cut with a matched pair of staff-bead and sunk-bead cutters, can be glued to construct wide panels. In addition, they are useful for making dust-proof hinged joints for display-cabinet doors. Both halves of the joint are cut in the same way, whether with a hand-held router or on an inverted-router table.

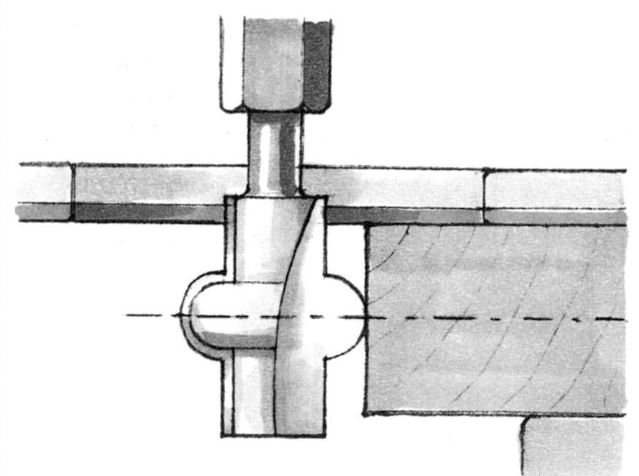

1 Setting up for hand routing
Clamp the board face-down, with the edge just overhanging the bench. Fit the sunk-bead cutter in the router and centre the bead on the edge of the timber, using the fine-depth adjuster, if fitted (see page 31).

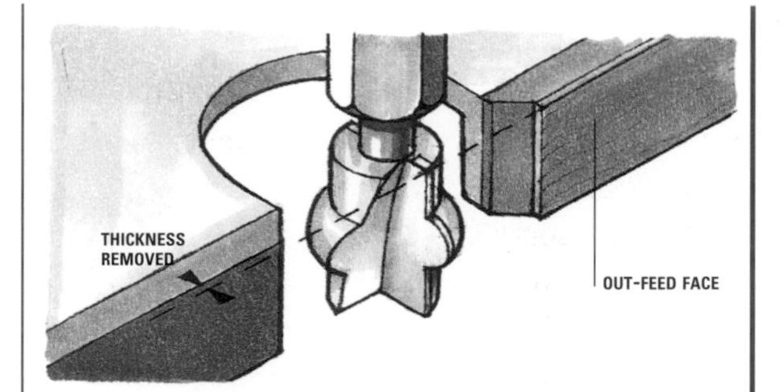

2 Fitting a scribed out-feed face
Set the side fence to leave the straight cutting sections of the cutter fractionally proud of the straight fence faces. Since the operation removes material from the full width of the guide edge, stick a strip of veneer to the out-feed face.

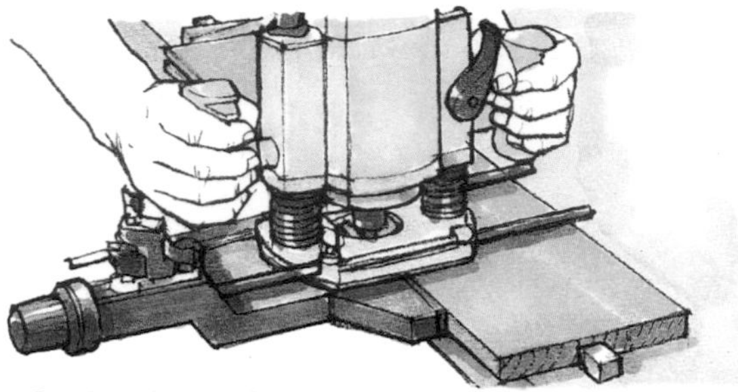

3 Cutting the profile
Feeding the router against the rotation of the cutter, machine the edge in a single pass, unless the wood is particularly hard or the grain is troublesome. Change the cutter and fit a shaped face (see page 46) to the out-feed fence, to machine the other half of the joint.

CUTTING A COVE JOINT

The cove joint enables you to join boards edge-to-edge to construct a curved panel. Both edge profiles are cut in a similar way (see left and above), but using shaped out-feed faces to run against the moulded edges behind the cutters (see page 46).

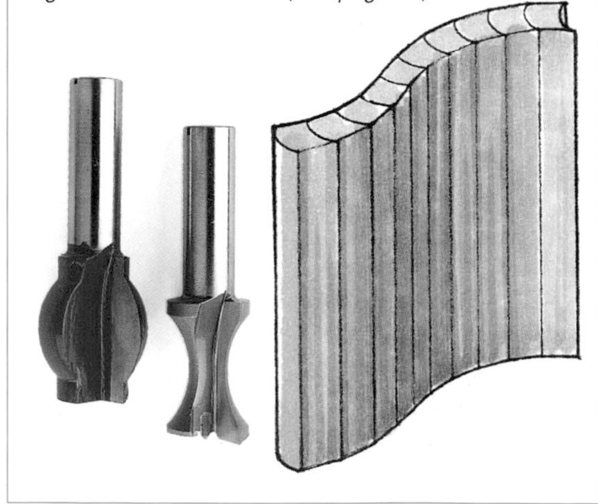

RULE JOINTS

A rule joint is not a woodworking joint in the conventional sense, since it does not physically attach one component to another. The joint comprises two moulded edges, one on a fixed table top and the other on a hinged leaf, designed to pivot one around the other as the leaf is raised and lowered. Its function is to conceal special back-flap hinges screwed to the undersides of the top and leaf. This type of hinge has flaps of different length, the longer one being screwed to the leaf. You can use a routing table to machine both edge profiles, but cut large tops and flaps on a bench, using a hand-held router.

Before you machine the actual table top and folding leaves, make a trial piece to check the fit of the rule joints, making a note of the settings.

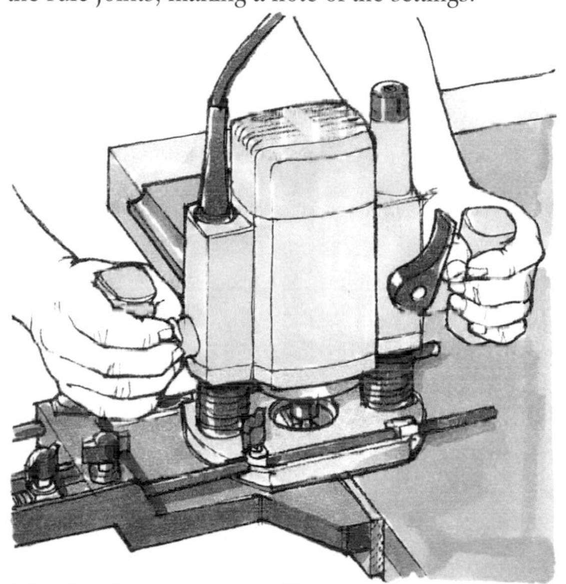

1 Cutting the concave profile
Clamp the leaf face-down on the bench. Fit a plain cove cutter in the router and set the depth of cut to leave a suitable shoulder along the top edge. This shoulder not only forms the guide edge for the fence or guide bearing, but also forms the visible, and most vulnerable, edge on the flap.

Set the fence in line with, or very slightly forward of, the axis of the cutter.

Set the depth of cut using the turret stop, finishing slightly less than the full depth. This gives you the opportunity to make a very fine cut on the final pass. Gradually lowering the cutter between passes, machine the profile against the rotation of the cutter.

2 Cutting the convex profile
To machine the fixed table top, clamp it face-up, fit a rounding-over cutter in the router, and set the full depth of cut to match the previously cut profile.

CHOOSING CUTTERS

Rule-joint sets are available for timber, ranging from 9mm (3/8in) to 18mm (3/4in) minimum thickness. Each set comprises one rounding-over cutter and a matching cove cutter. Fit a side fence with extended faces to guide a set of plain cutters. Run the bearings of self-guiding cutters along the uncut shoulder of each profile. If necessary, clamp a batten flush with the guide edge to provide sufficient running surface for the bearing.

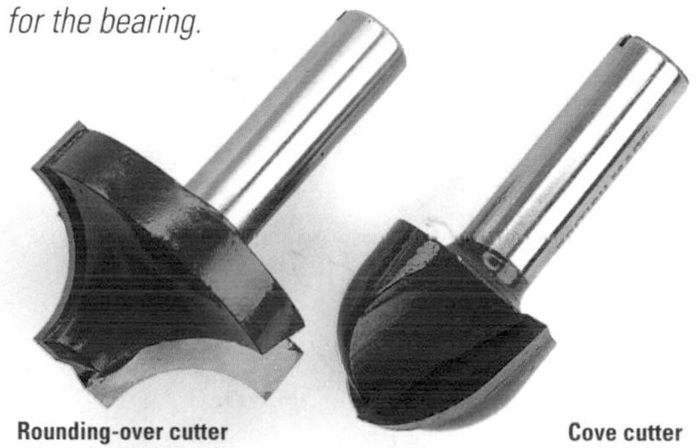

Rounding-over cutter **Cove cutter**

SCARF JOINTS

Scarf, or splicing, joints are used traditionally for joining two lengths of solid timber end-to-end. Scarf joints can be cut using a router, with the aid of a simple jig. The degree of taper should be between 8:1 and 12:1, the latter being used for work where the joint is likely to be stressed.

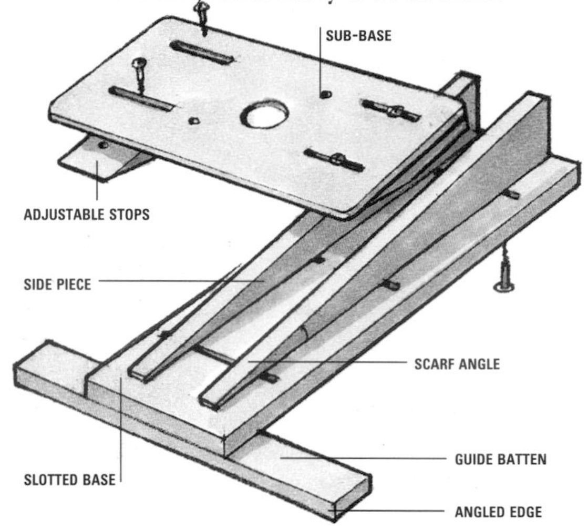

Making the jig

Cut two tapered side pieces from 18mm (¾in) thick timber to the scarf angle, but deeper than the height of the workpiece. Screw these to a slotted base plate at a distance apart equal to the width of the workpiece. A relatively wide base plate allows enough room for clamping and can be also used for wider projects. Glue a guide batten flush with the end of the base plate, and plane the batten square to the taper angle.

Cut a wide sub-base for the router, fitting two stops on the underside to limit sideways travel across the jig. Set these so that the router will cut into the inside faces of the tapered side pieces by about 2mm (³⁄₃₂in).

Roughing out the workpiece

To save time and cutter wear, first saw the workpiece by hand or machine roughly to the taper angle; the router will finish it cleanly and precisely.

Cutting a plain scarf joint

Work on a bench that is large enough to support both the jig and workpiece. Clamp the jig to the workpiece, with the rough-sawn face set level with the tapered edges of the jig. Choose a large-diameter cutter with a good bottom-cutting facility, and set the depth of cut below the sub-base to about 3mm (⅛in).

Machine the length of the scarf in a series of straight cuts. Adjust the depth of cut until a feather edge is produced at the bottom of the taper and the face is flat and level.

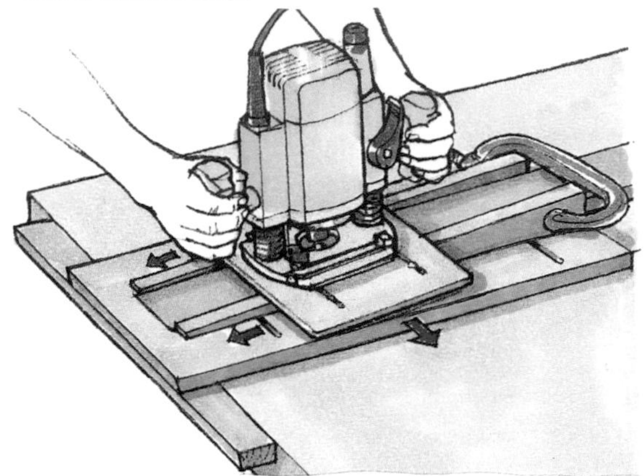

Cutting a hooked scarf joint

For more positive location without feathered edges, make a hooked scarf joint. Fit a deep face to the side fence, and set it to cut the top shoulder of the joint. Cut to the required shoulder depth before machining the tapered face of the scarf as before. Take care not to cut into the shoulder. After you have cut the face, re-set the side fence to cut across the tip of the joint.

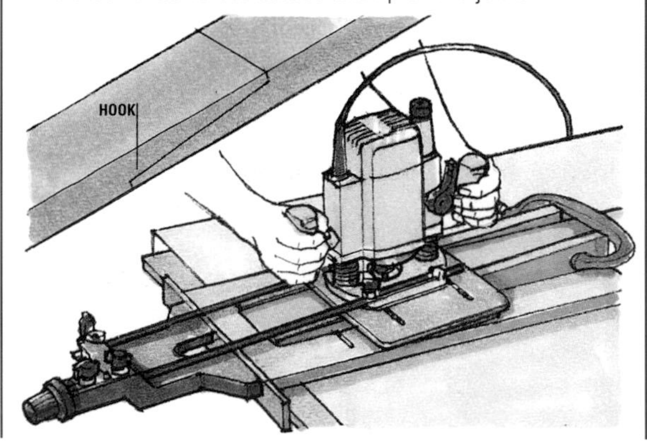

HALF-BLIND LAP JOINT

The half-blind lap joint is sometimes used by woodworkers as a substitute for the lapped dovetail in drawer construction. A lap joint is not nearly as strong as a dovetail, but it saves you the expense of a special factory-made jig. When making the joint, prepare and sand the wood beforehand, since any change in thickness afterwards will result in a loose joint.

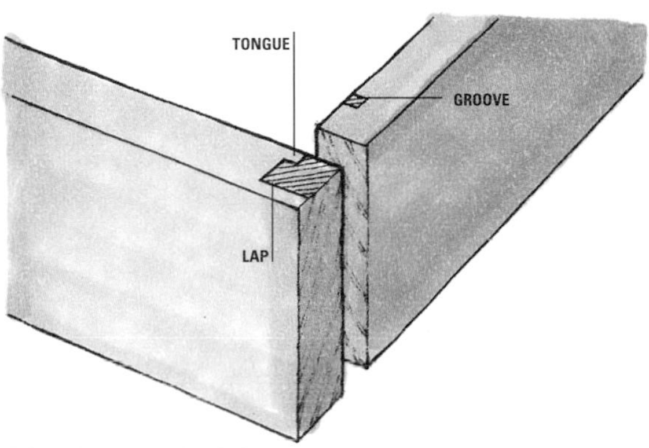

Marking out the joint
In order to visualize the joint and help you set up the router, first mark out the joint on one pair of components. Other identical joints will not require marking. The lap on the rabbet member should be about 4 to 6mm (5⁄32 to 1⁄4in) thick. Choose a router cutter of a similar diameter to machine the groove in the side member, and plan to cut it to a similar depth.

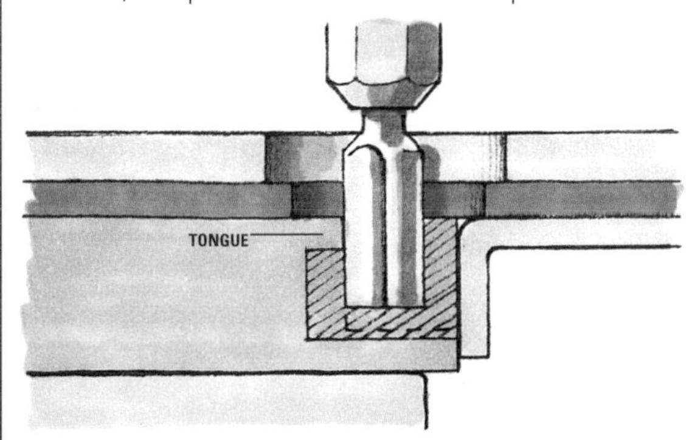

1 Machining the rabbet member – first cut
Clamp the workpiece face-down on the bench. Adjust the router fence until the cutter will remove the waste up to the end of the tongue. Remove the waste in stages, stopping just short of the lap. You can machine several identical components at once if you clamp them side by side, flat on the bench.

2 Machining the rabbet member – second cut
Clamp the work upright in a vice. To help steady the router, clamp a guide batten to the face side, flush with the end grain. Adjust the router fence to trim the lap to the required thickness, and set the depth of cut to remove the remaining waste. Re-set the fence to make another pass with the cutter, leaving the tongue at the required thickness.

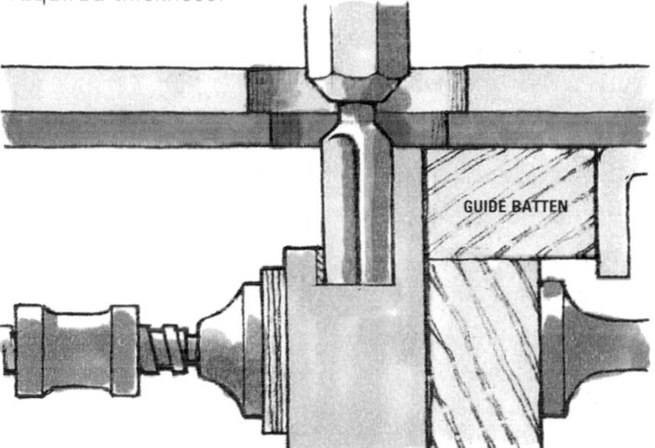

3 Cutting the groove
Set the depth of cut to match the length of the tongue, and adjust the fence to place the groove the required distance from the end of the workpiece. Make the cut with a single pass. Once again, you can save time by machining several workpieces at once.

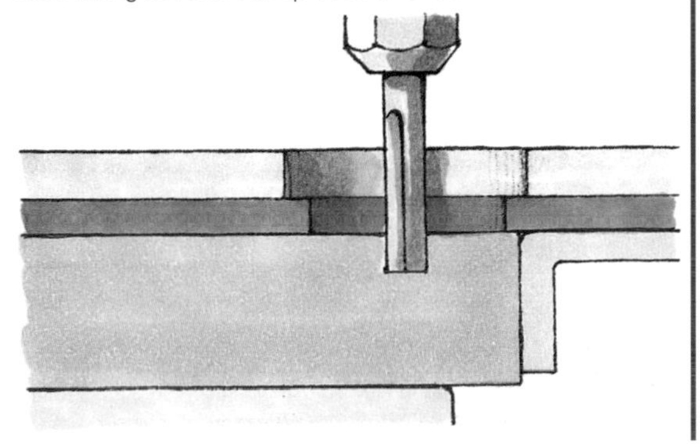

DOVETAIL JOINTS

Dovetail joints can be easily and quickly cut using special factory-made jigs and a power router fitted with a guide bush and cutter that ensure the size and pitch of the pin and tail match exactly (see page 25). Some jigs can be modified to cut a number of different joints, but the descriptions that follow serve to explain the principles of cutting the two most commonly used dovetail joints, the through and lapped versions. The former, which are visible on both sides of a corner, are used in cabinet construction and for the rear joints in drawers. Lapped dovetails, often used for attaching drawer fronts, cannot be seen when the drawer is closed.

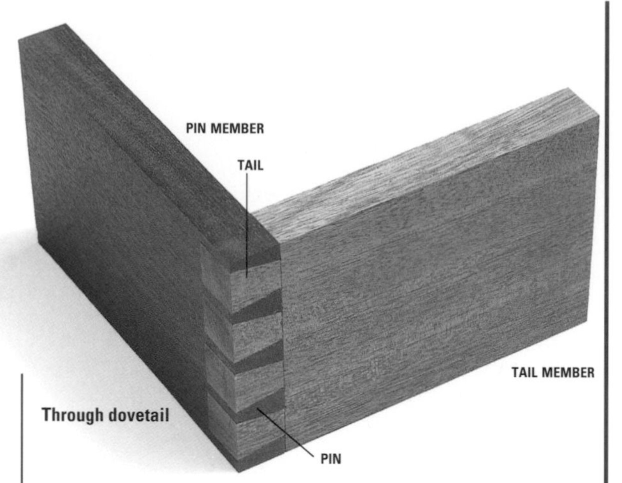

Through dovetail

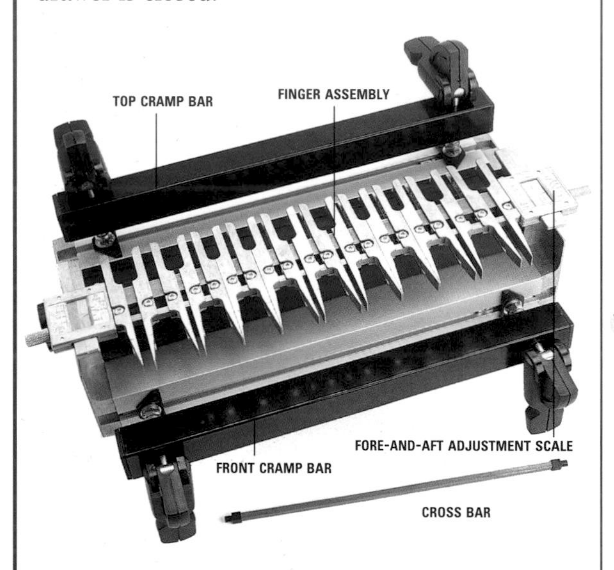

Using an adjustable-finger jig

You can tailor a through dovetail joint to suit any width of workpiece, using a jig that features individually adjustable fingers for guiding the router cutter. The setting of the finger assembly governs the size and spacing of the tails and pins. Adjusting the fingers for the pins along one side of the assembly automatically sets the fingers for matching tails on the opposite side. There is also a device for fine fore-and-aft adjustment of the finger assembly.

Fitting a test piece

Use test pieces to set up the jig and check the fit of the parts. Place an 18mm (¾in) spacer board under the finger assembly and secure it with the jig's top cramp bar. Clamp a test piece of the correct width to the front of the jig, making sure it butts against the side stop and the underside of the finger assembly.

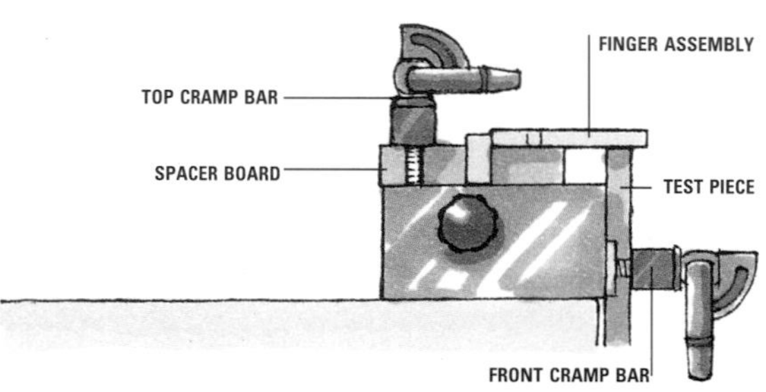

Adjusting the finger assembly

You need one pair of fingers for each dovetail and a single finger at both ends of the row. Set these individual fingers first so that they are flush with each edge of the test piece, then space the pairs of fingers at regular intervals between them. Tighten the clamping screws.

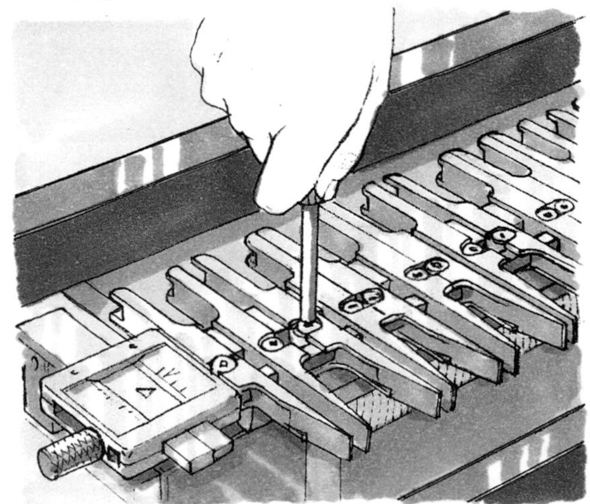

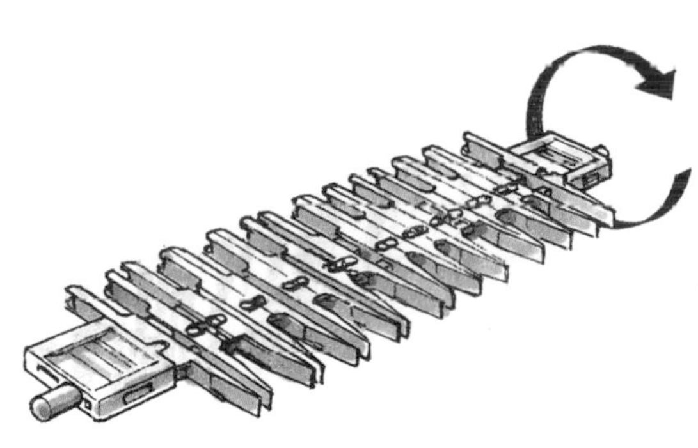

Rotating the fingers
Once you have adjusted one half of the assembly, all that is required to utilize the other set of fingers is to unclamp the entire assembly, rotate it and clamp it down again.

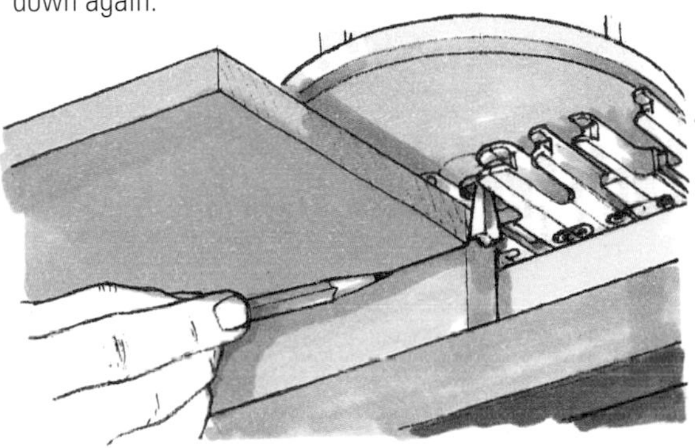

1 Setting up for tail-cutting
Set the assembly with the straight-sided tail fingers pointing towards you. Fit a compatible dovetail cutter and guide bush into the router – check with your jig supplier. Using the other component as a template, mark the depth of cut on the tail member, and adjust the router cutter down to the marked line.

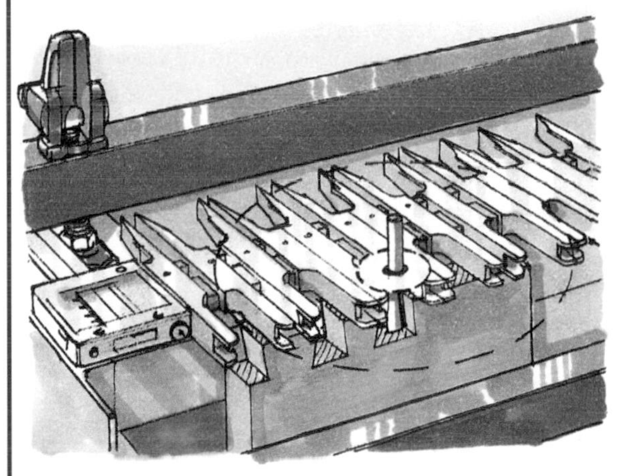

2 Cutting the tails
Always check that all parts of the jig and the test piece are securely clamped before switching on the router. Then run the cutter between each set of fingers, cutting out the waste wood and leaving a neat row of tails.

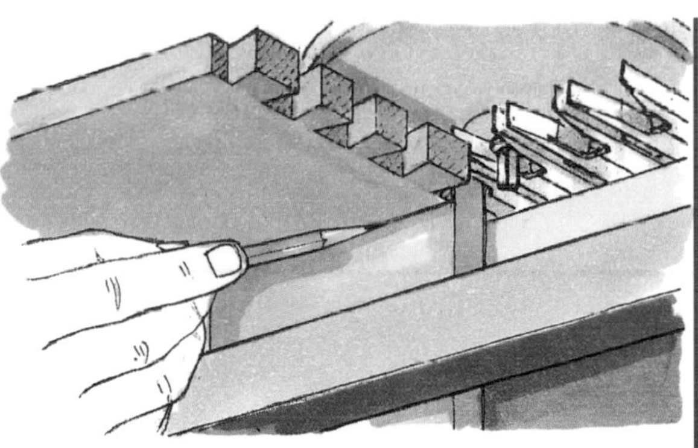

3 Setting up for pin-cutting
Rotate the assembly to present the pin fingers to the front of the jig, and clamp a second test piece in the jig. Mark the depth of cut as before, using the tail member you have just cut as a template. Fit a straight cutter in the router and adjust it down to the marked line.

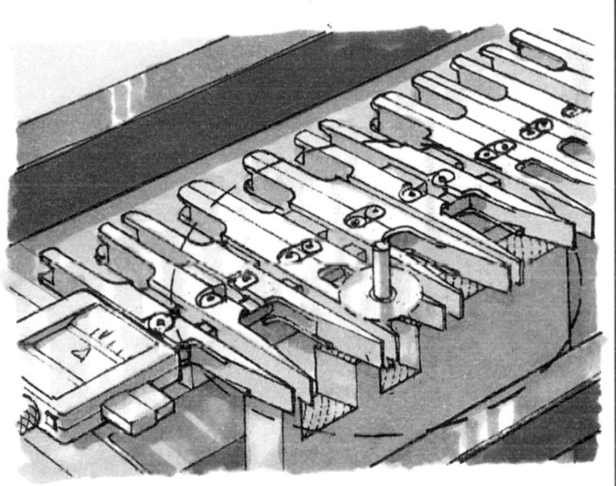

4 Cutting the pins
With the router placed flat on the finger assembly, switch on and then rout out the waste between each pair of fingers.

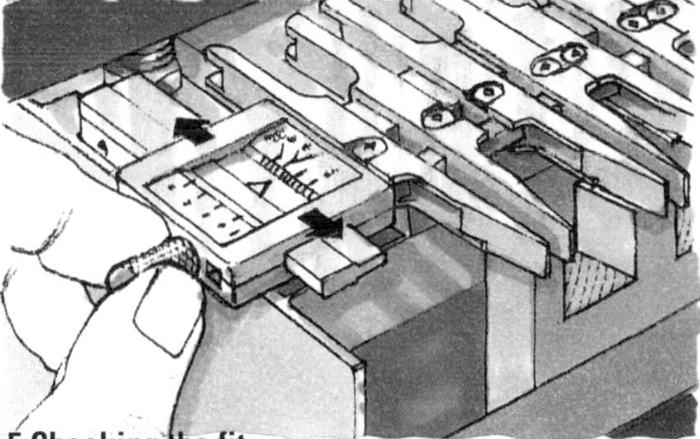

5 Checking the fit
Assemble the joint to check it fits snugly. If it is slack, use the jig's fine-adjustment scale to move the finger assembly forward. If the joint is too tight, adjust the finger assembly backwards. Cut another pair of test pieces to check that the joint fits satisfactorily, before proceeding with the actual workpieces.

USING A SLIDING-CARRIAGE WORKCENTRE

You can cut virtually any type of dovetail joint on a sliding-carriage workcentre (see page 86). However, the methods shown here are for cutting a simple through dovetail. The difference between using this jig and the finger-template jigs is that the control and clamping mechanisms allow the pin and tail spacings to be set precisely by eye. Although each pin member is cut separately, several tail members can be cut at the same time.

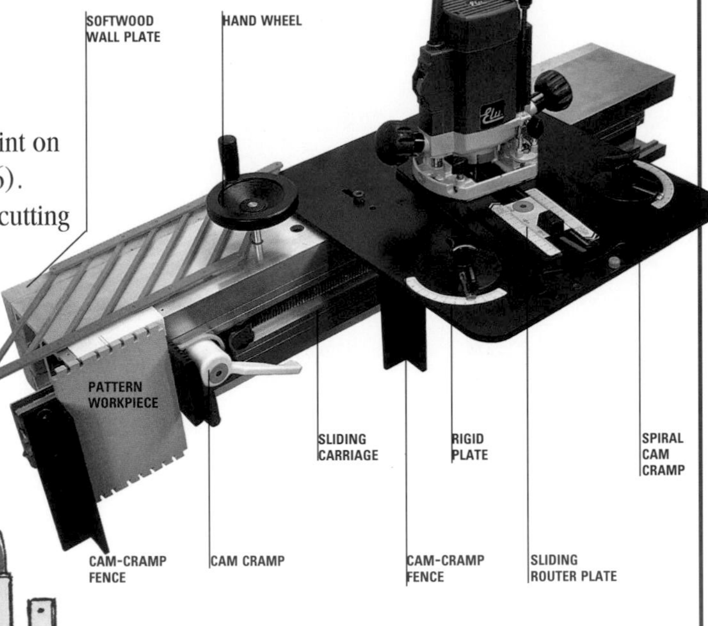

Setting up the workcentre

Any dovetail cutter can be used, as there is no specific template or guide bush to be matched. Set the spiral cam cramps to guide the router plate square into the workcentre. Set the depth of cut to suit the thickness of the workpieces. Clamp a test piece face-on in the right-hand cam-operated cramp.

Cutting a socket

Cut a dovetail socket by pushing the cutter through from the front face – a block of wood is fitted in the workcentre to prevent break out on the back face. Relocate the test piece in the left-hand cam-operated cramp and, using a sharp pencil, trace the shape of the socket onto a white label stuck to the face of the workcentre. Remove the test piece.

Setting out the tails

Lock the pin member in the left-hand cam-operated cramp and the tail member in the right-hand cramp. Both workpieces move together, the left hand for marking out, and the other piece set up to be cut. Mark out the required pin positions along the top edge of the pin member, using the drawn socket as a guide.

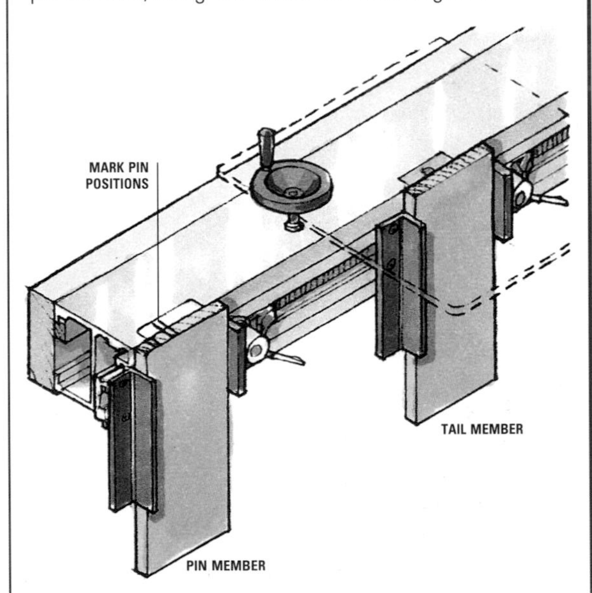

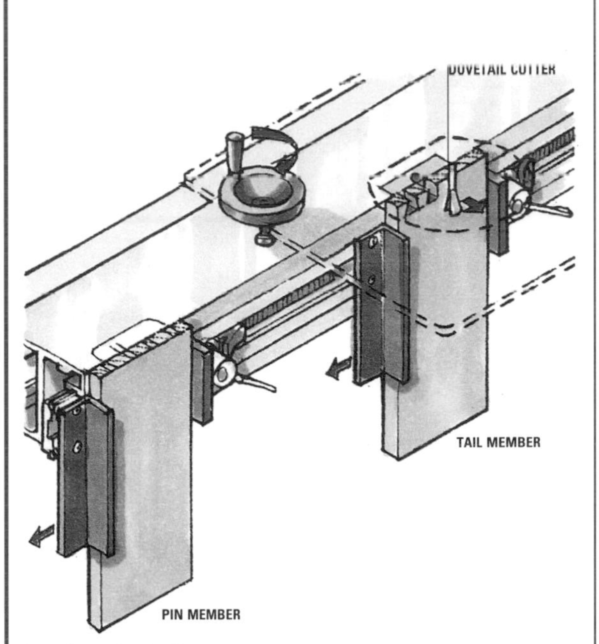

Cutting the tails

Wind the sliding carriage complete with the two workpieces by turning the hand wheel on top of the workcentre to align each marked pin position with the drawn socket. Each time, cut the corresponding dovetail in the tail member.

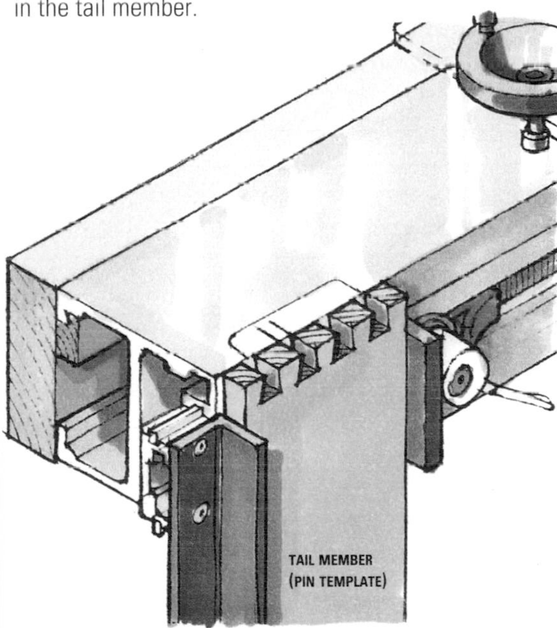

Clamping the pin template

The tail member now becomes the template for cutting the pins. Fix the pin member in the right-hand cam-operated cramp. Fit a straight cutter of a diameter smaller than the narrow end of the cut dovetails, and set the depth of cut to suit the workpieces. Set the pivot point to allow the router to cut at an angle.

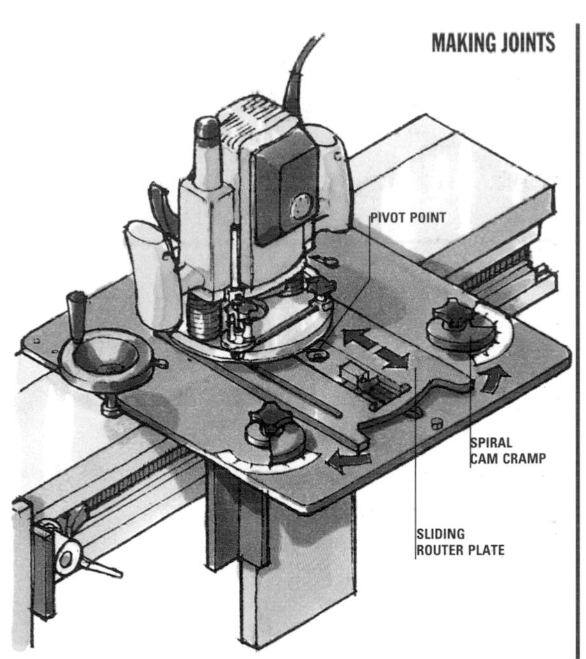

Setting up the router plate

Set the router plate to cut at the required angle across the rigid plate, within the limits set by the spiral cramps and pivot point. Refer to the manufacturer's instructions for determining angles. Turn the spiral cramps on the rigid plate to the required angle, and tighten them.

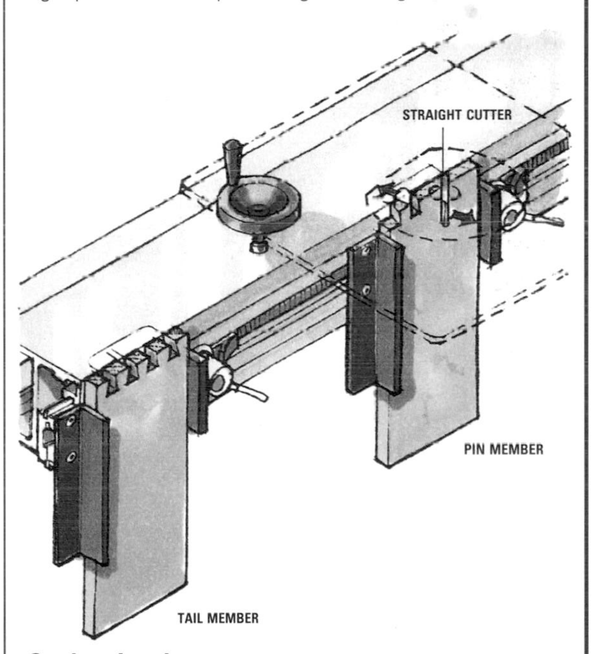

Cutting the pins

Align each cut socket on the tail member with the drawn socket, and cut the pins accordingly. For this, the router cutter is first passed through the pin member, with the router plate held against one spiral cramp and then against the other, to cut both angled faces of each pin. Remove the waste from between the pins.

LAPPED DOVETAIL JOINTS

A lapped dovetail joint can be cut with a hand-held power router, using a fixed-finger dovetail jig that enables both the pins and tails to be cut simultaneously. It is a relatively inexpensive jig that produces equal-size, regularly spaced pins and tails. The joint is perfectly functional, but it is necessary to design the width of the components to suit the finger spacing of the jig. Test the jig's settings by cutting test pieces before you proceed with the actual work.

1 Clamping the work
Mount the tail member vertically in the jig, face-side inward, then insert the pin member (drawer front) face-down, and butt its end grain against the tail member. Slide the pin member up to the jig's edge stop, then offset the tail member sideways by half the finger spacing. Now fit the finger template, which is marked with a 'sight line' that runs centrally down the row of fingers. Adjust it until the sight line corresponds to the butt joint between the two components.

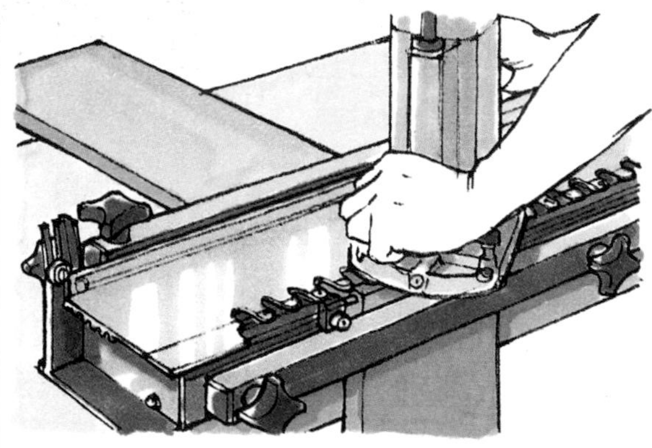

2 Cutting the joint
Fit the recommended guide bush and dovetail cutter. Make a light back cut from right to left, to leave a clean shoulder, then feed the cutter from left to right between each pair of fingers, keeping the router level and following the template with the guide bush.

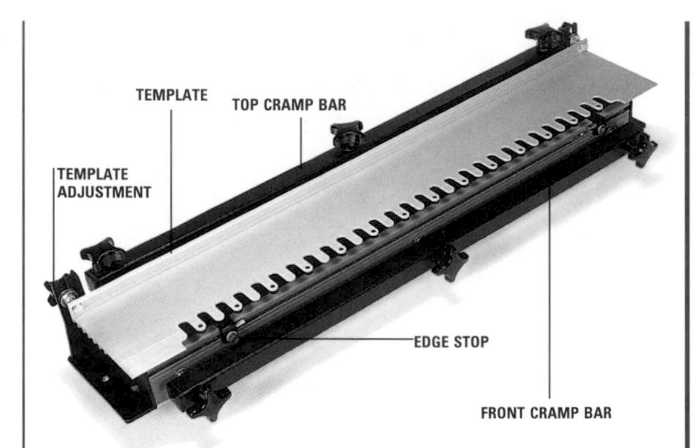

3 Assembling the joint
Unclamp the test pieces and rotate one of them through 180 degrees to mate their jointed ends. If the joint fits snugly, cut a similar joint for the other end of the pin member (drawer front), butting it against the edge stop at the other end of the jig.

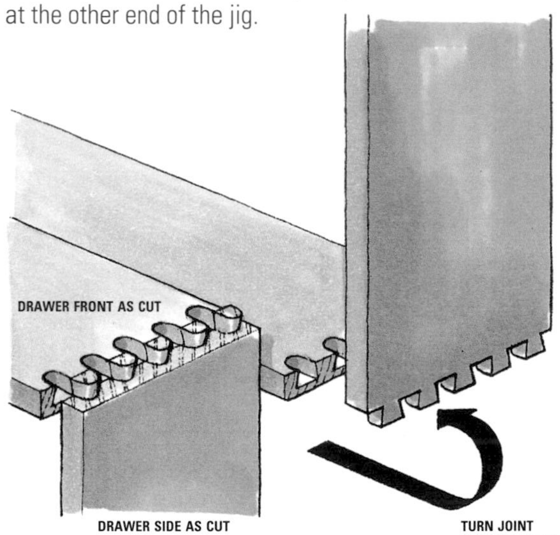

4 Adjusting the cut
If you find the joint is loose, increase the cutting depth of the router slightly. If the joint is too tight, raise the cutter.

5 Modifying sockets
If the sockets are too deep, adjust the finger template forward. If the tail member projects slightly, set the template backwards.

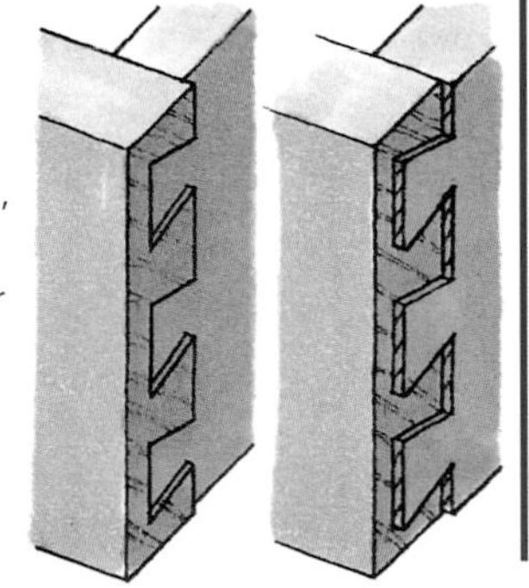

CHAPTER 9 Developing skills and techniques may be a source of satisfaction in itself, but every woodworker wants to make things he or she can use. The following pages set out ways of making simple unit furniture that incorporate basic and advanced techniques, demonstrating the potential for making accurate and cleanly cut components with the power router.

BASE UNITS

Maximum use and economy of space can be provided by built-in furniture, using base units not only as cupboards and drawers, but also to house appliances and to support a continuous worktop. Base units can be fitted with full-height doors or can incorporate sets of drawers, or a combination of both of these.

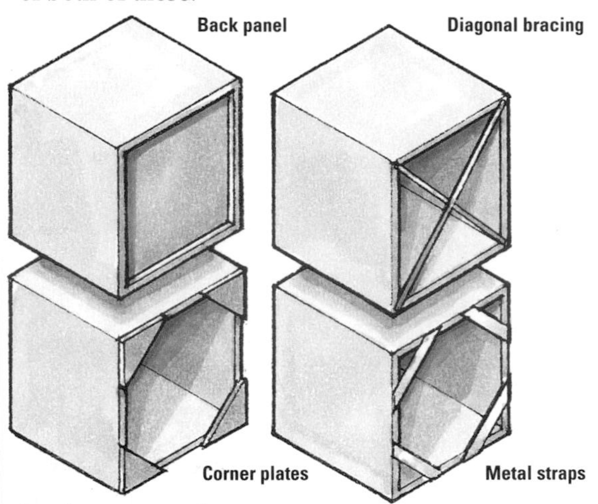

Basic construction

The simplest method of construction comprises two sides, a top and bottom, and a back panel. Although the cabinet is held together by corner joints, the back panel adds diagonal stiffness to the structure. Where there is no need for a back panel, use diagonal braces or corner plates to brace the structure.

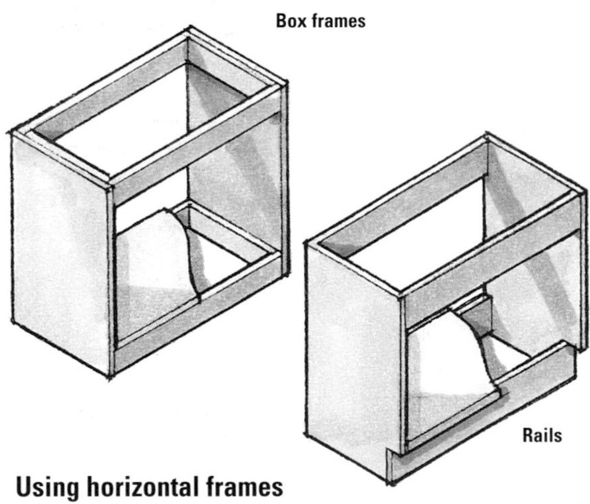

Using horizontal frames

An alternative way to stiffen a carcass (the cabinet box) is to join the two end panels using shallow box frames at the top and bottom. This can be simplified down to deep front and rear rails, with the former set back at the bottom to form a plinth board.

STANDARD SIZES AND PROPORTIONS

Base units are designed to support a worktop and accommodate standard-size, or 'modular', appliances. A typical appliance is 600mm (2ft) deep and 900mm (3ft) high. Manufactured units are made to standard widths, to enable a continuous run of cabinets to be fitted into a given space. Single units are generally 300, 400, 500 or 600mm (12in, 1ft 4in, 1ft 8in, and 2ft) wide, and double units are made 800mm and 1m (2ft 8in and 3ft 3in) wide. A recessed plinth provides toe clearance for someone standing at the worktop.

Adopting standard widths when you construct home-made units gives you the option of fitting ready-made, store-bought doors.

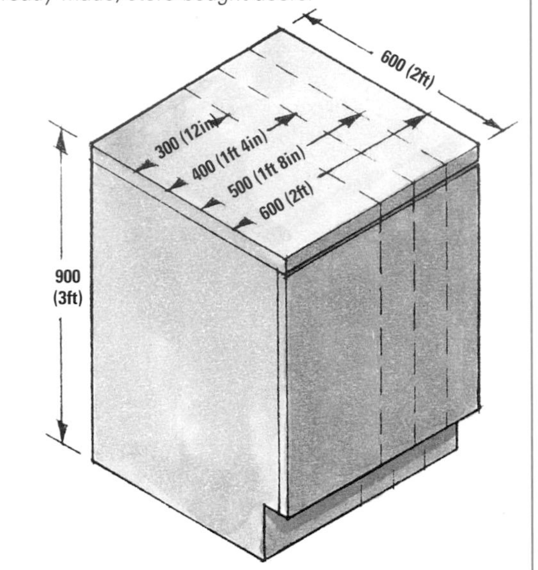

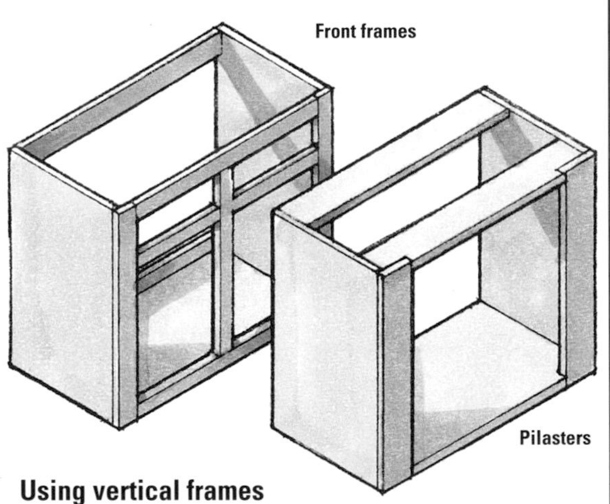

Using vertical frames

Vertical front frames can be used to stiffen a carcass and to provide a centre support for doors and drawers. You can then use rails and dividers to impose a different style on the cabinet. Wide vertical pilasters perform a similar stiffening function.

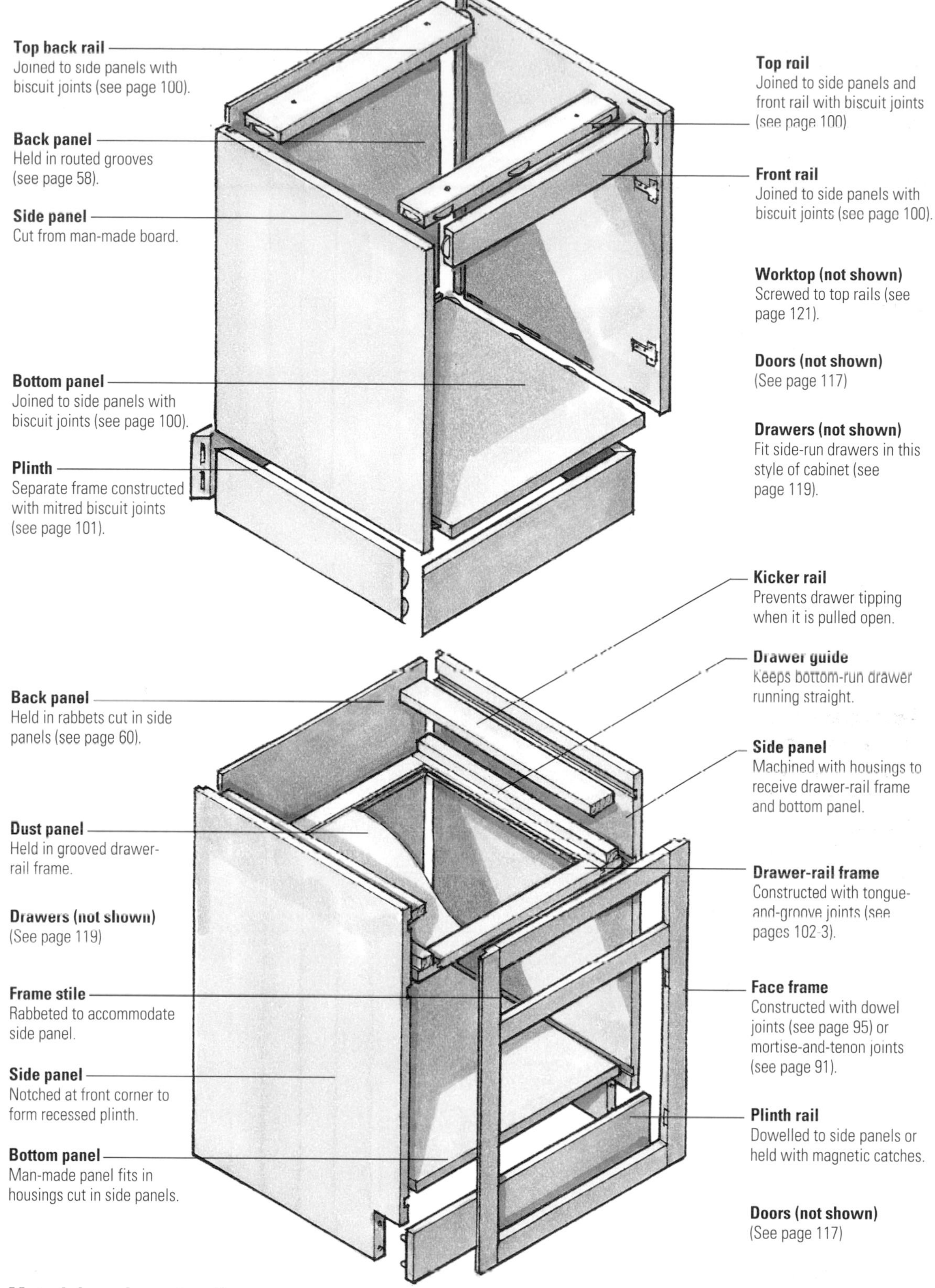

Materials and construction

Most units are constructed from veneered or melamine-faced chipboard, a durable, solid and stable material.

Biscuit dowels can be used for both solid-timber and veneered or melamine-face construction (see page 100). With the latter, however, skim away the thin surface film along the centre line of each joint, using the tip of a 10mm (3⁄8in) straight cutter; this exposes the board surface and allows the glue to bond.

WALL-HUNG UNITS

Install wall units to further increase storage capacity, particularly in a kitchen or workroom. These can be constructed in the same fashion as the base units, but always with a solid top panel. Methods for hanging wall units include proprietary fixings and home-made bevel-edge battens.

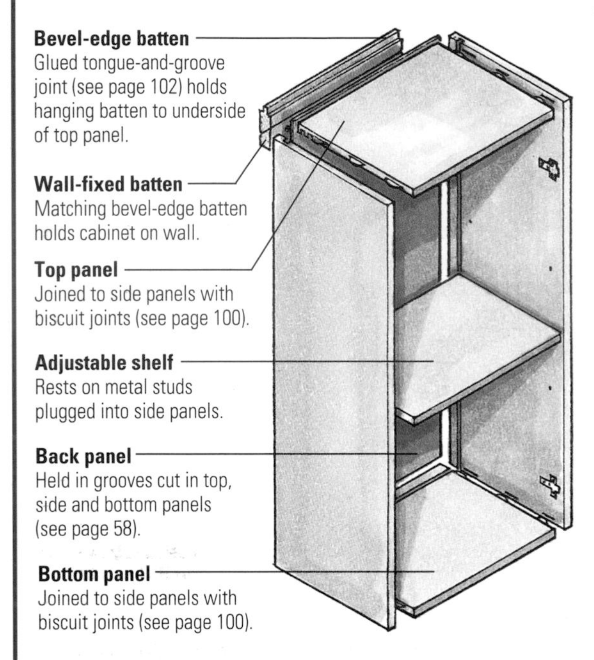

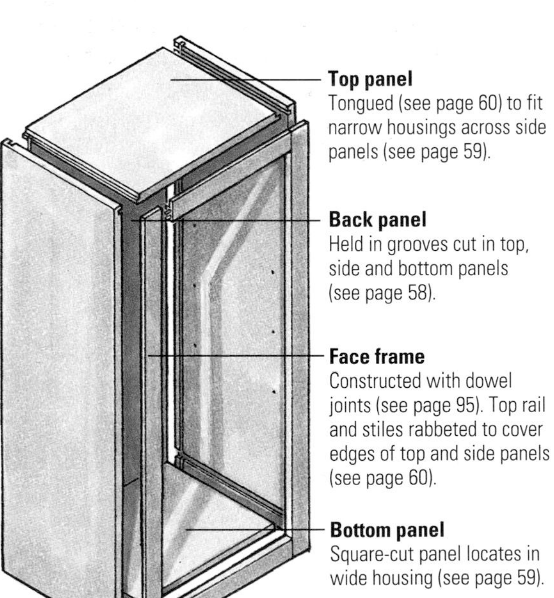

Wall-unit proportions
Proprietary wall units are made to the same widths as base units, but are invariably 300mm (12in) deep. Although it may mean that only the bottom section is easily accessible, you can make wall units to suit any ceiling height, either as a close fit or by leaving a gap. A router-cut cornice moulding can be fitted to trim the top edge or used to fill a narrow gap.

SHELVES

Open shelves can be made to match wall units, using similar materials and construction techniques such as biscuit jointing. Alternatively, make up shelves from solid timber, using square-cut or dovetail housings (see page 78).

If you make up hanging-shelf units to similar dimensions as wall units, they can be used to break up a run of cupboards and introduce display space for books, china and other artefacts.

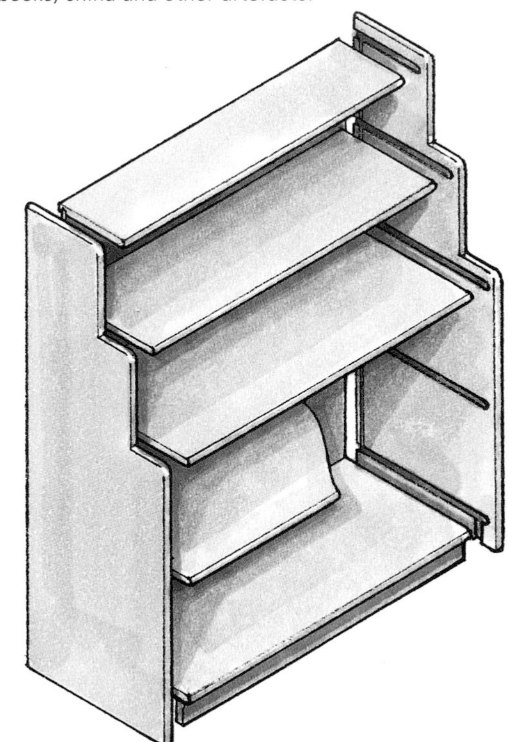

Floor-standing shelves
Construct floor-standing shelf units on similar lines. 'Waterfall'-style units, where the shelves are wider towards the bottom of the unit, allow the base depth to be increased for better stability. Use moulding cutters to machine decorative edges (see page 61).

CABINET DOORS

You can use a router not only to cut decorative mouldings on the face and around the edges of cabinet doors, but also to construct traditional framed doors with raised-and-fielded panels.

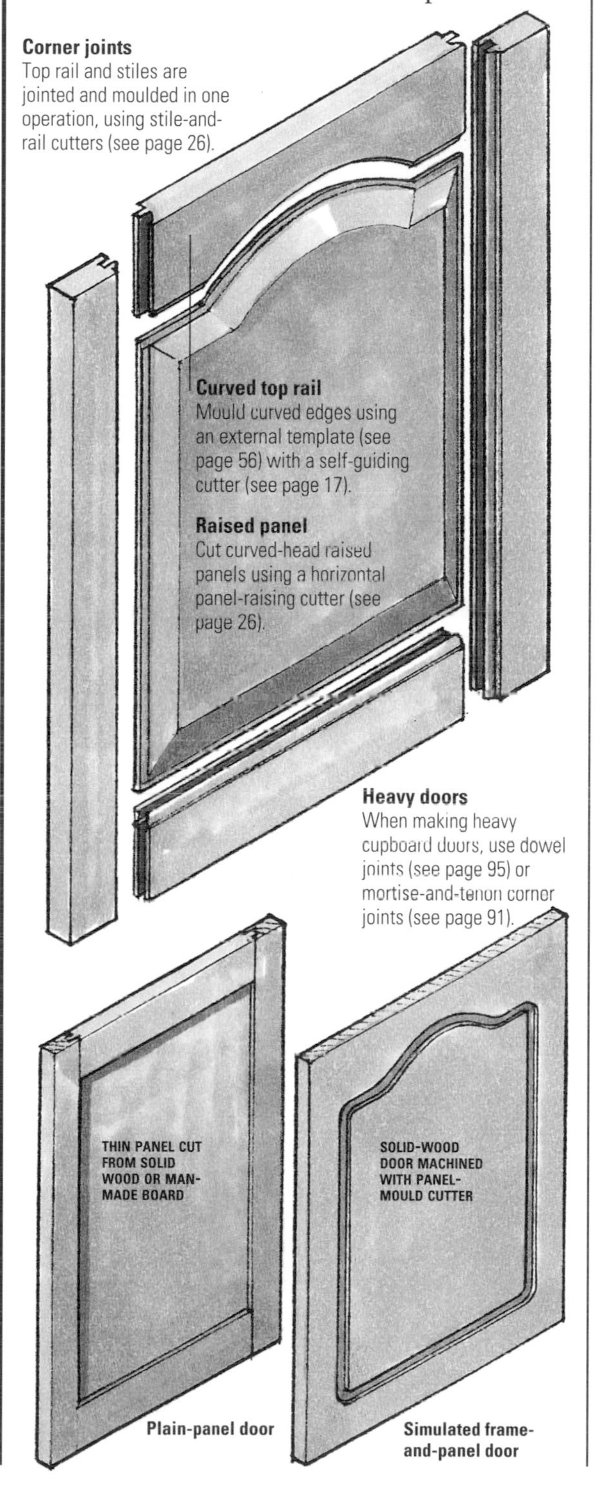

Corner joints
Top rail and stiles are jointed and moulded in one operation, using stile-and-rail cutters (see page 26).

Curved top rail
Mould curved edges using an external template (see page 56) with a self-guiding cutter (see page 17).

Raised panel
Cut curved-head raised panels using a horizontal panel-raising cutter (see page 26).

Heavy doors
When making heavy cupboard doors, use dowel joints (see page 95) or mortise-and-tenon corner joints (see page 91).

Plain-panel door

Simulated frame-and-panel door

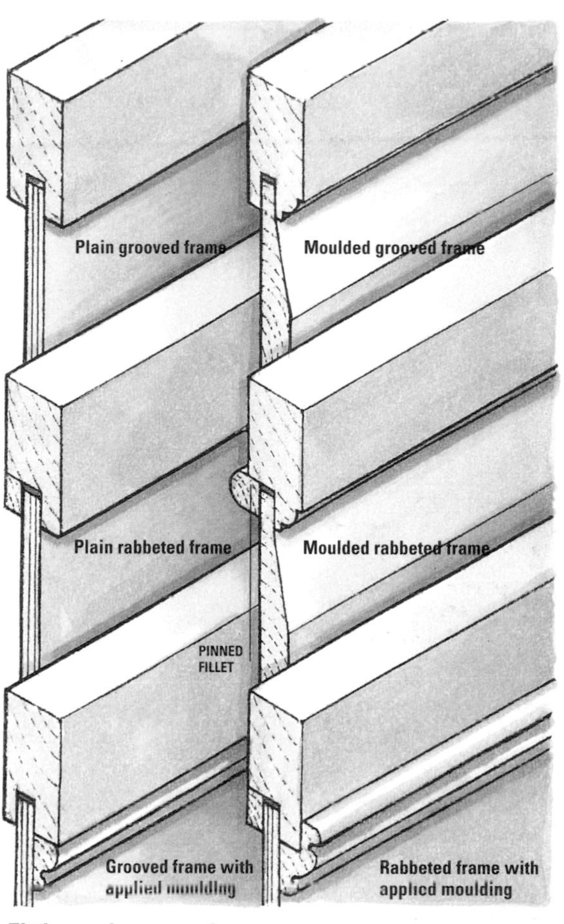

Fitting a door panel
Fit a plain or raised panel into a groove routed along the inside edges of door rails and stiles. Alternatively, cut a rabbet and fix the panel with pinned fillets. In either case, do not glue the panel in place. For decoration, mould the edges of the frame or use a separate applied moulding.

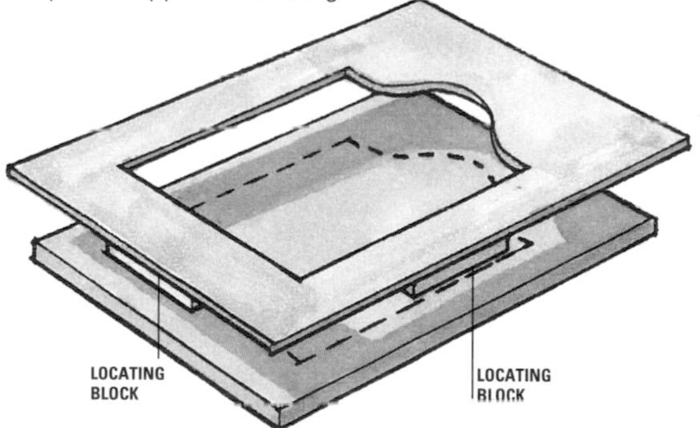

Making a face-moulding template
Cut an internal template from 6mm (¼in) MDF, leaving a generous margin all round for clamping to the work. Fit locating blocks on the underside, to fit over the edges of the workpiece.

HANGING DOORS

Most cabinet doors are hung using flap hinges set into a rectangular recess, or with concealed hinges screwed to the jamb and set into a round recess on the back of the door.

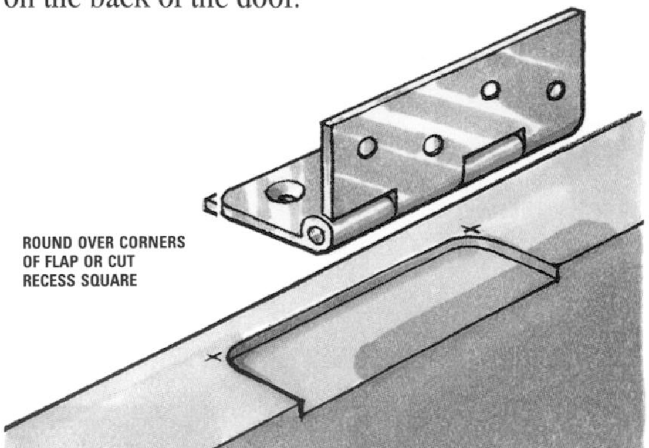

Flap-hinge recesses
Cut rectangular recesses using a simple template/guide bush combination. Use a small-diameter, two-flute straight cutter for small cabinet hinges. The cutter will leave rounded corners that will need to be finished with a sharp chisel. Alternatively, you can round over the corners of the hinge flap with a fine file.

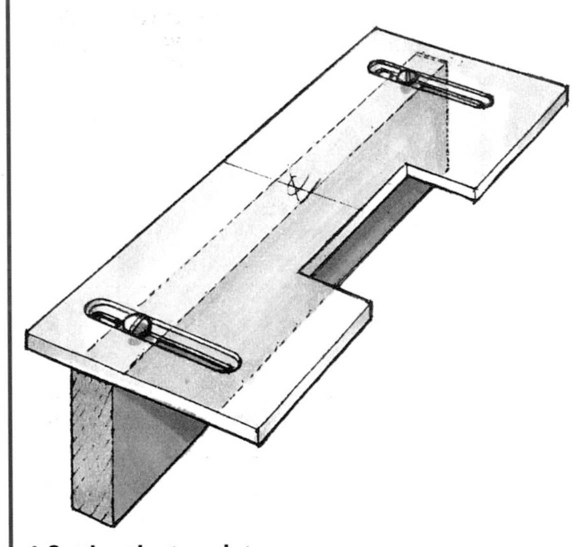

1 Cutting the template
Cut the template as an open-side rectangular cutout in 6mm (¼in) clear plastic. Allow adequate area to balance the router, and allow for the guide-bush margin (see page 50). Cut two short slots across the width at either end, and recess them to take two pan-head screws. Fit a clamping batten to the underside; the two clamping screws give adjustment across the door or jamb width.

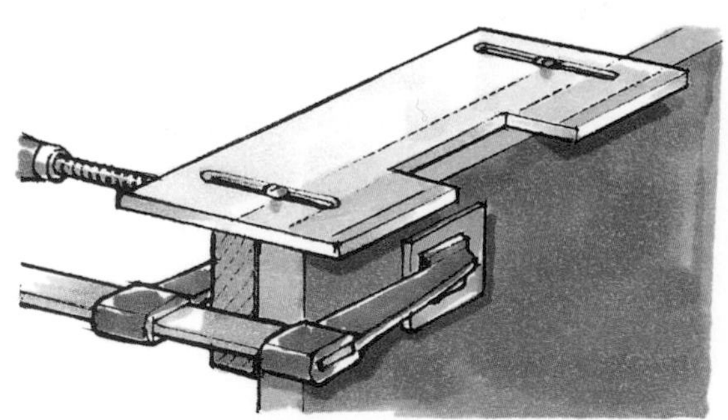

2 Cutting the recesses
Clamp the template to the edge of the door or jamb. Set the depth of cut by adjusting the router depth stop to the thickness of the hinge flap, allowing for the thickness of the template. Set each recess an equal distance from the top and bottom of the door, and align the end of the template against the edge of the door or against a marked line. When cutting the jamb recesses, clamp the template to the carcass, making sure that the router is not impeded by the cramp.

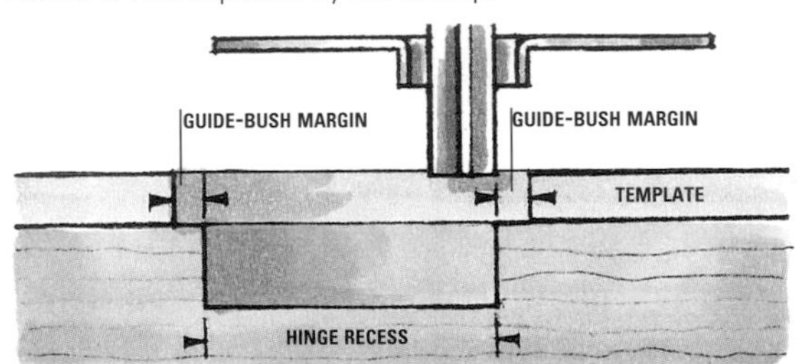

1 Cutting round hinge recesses
Hinge-sinking cutters are available to cut round recesses, but these must only be used at slow speeds in a drill press or a fixed router. To form round recesses using a hand-held router, make a template and choose a straight cutter with good bottom-cutting facilities, with a matching guide bush. Calculate the guide-bush margin, adding this to the hinge-recess diameter.

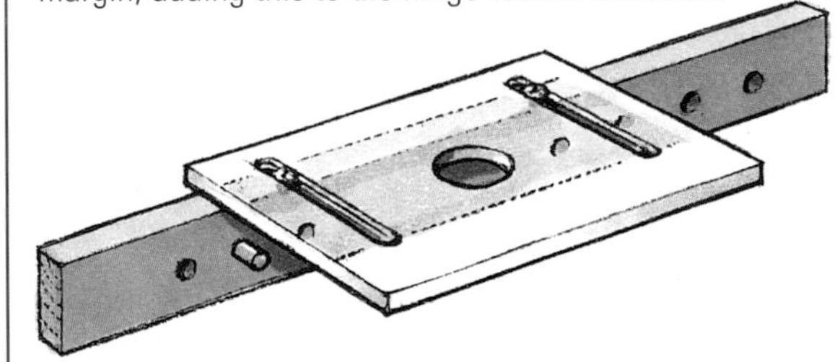

2 Making the template
Select a hole saw of the right diameter, and insert it in a power drill. Cut a circular recess in 6mm (¼in) thick clear plastic, and cut slots as for a flap-hinge template. Cut the clamping batten longer than the width of the template and drill holes at regular intervals, to take a loose dowel for setting the distance in from the top and bottom edges of the door.

DRAWERS

Although it is possible to make double-lapped and through dovetails, using special dovetail jigs (see pages 108-112), the router also lends itself to alternative methods of drawer construction, based on half-blind lap joints (see page 107) or the basic version (see right).

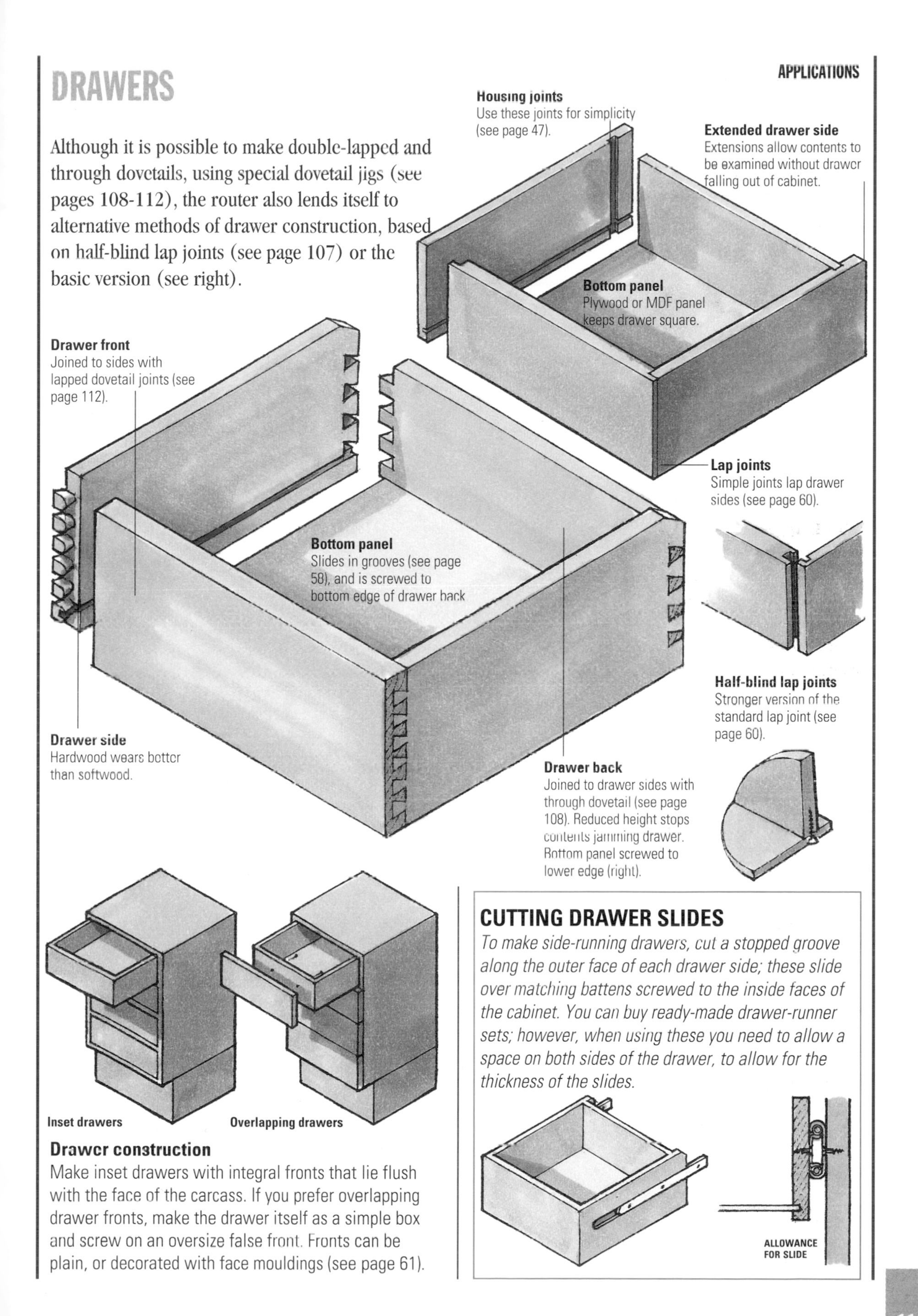

Drawer construction

Make inset drawers with integral fronts that lie flush with the face of the carcass. If you prefer overlapping drawer fronts, make the drawer itself as a simple box and screw on an oversize false front. Fronts can be plain, or decorated with face mouldings (see page 61).

CUTTING DRAWER SLIDES

To make side-running drawers, cut a stopped groove along the outer face of each drawer side; these slide over matching battens screwed to the inside faces of the cabinet. You can buy ready-made drawer-runner sets; however, when using these you need to allow a space on both sides of the drawer, to allow for the thickness of the slides.

WORKTOPS

The router is the ideal tool for trimming and cutting laminate-faced worktops. It is also extremely useful for forming cutouts and inset holes, since laminate-trimming cutters leave the edges perfectly clean and square.

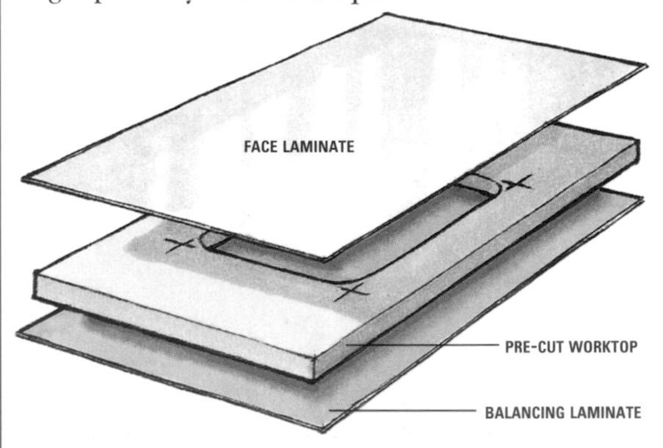

Working with laminates

Laminate-faced worktops are generally square-edged, requiring either an applied laminate or solid-timber edging, or they are manufactured with a rounded bullnose front edge. All laminate-faced materials must have a balancing laminate applied to the underside, to resist any tendency for the worktop to bow.

If you laminate a work surface yourself, make the cutouts for inset fittings before applying the face laminate. Use a pierce-and-trim cutter to follow the internal edge of the opening (see page 23).

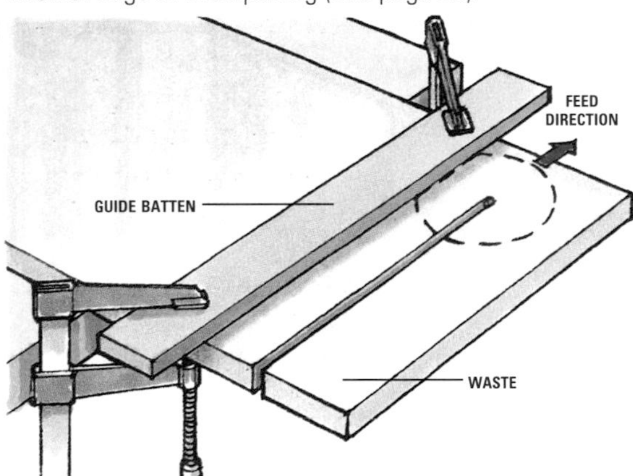

Cutting worktops to size

To cut a worktop to length, clamp a straightedge across the workpiece to guide the router base (see page 47). Always clamp the straightedge on the 'good' side of the cut line, so that any accidental wavering off course will run the cutter into waste material.

Fit the router's side fence when cutting a worktop lengthways, to make it narrower.

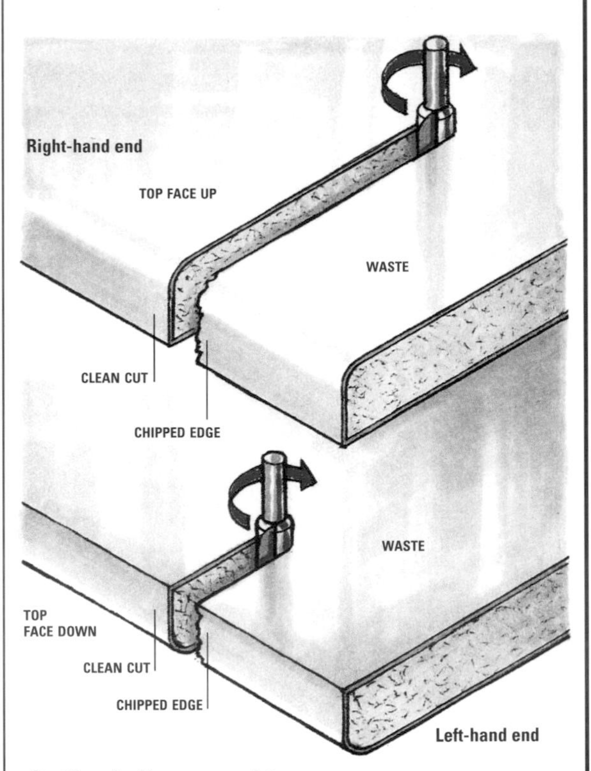

Cutting bullnose worktops

When cutting laminated bullnose worktops or pre-edged workpieces to length, cut a right-hand end with the laminate face-up, feeding the cutter from the front edge. Cut left-hand ends with the worktop clamped face-down, to prevent the laminate chipping.

MATCHING CUTTER TO ROUTER

Since ready-made worktops are relatively thick, it is best to use a heavy-duty router when severing them. You can use a medium-duty machine, but make the cut in a series of very shallow passes.

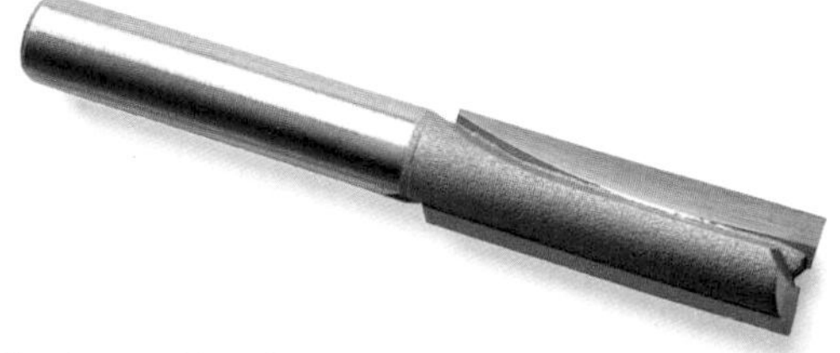

Router-cutter sizes

When using a heavy-duty router, fit a 12mm (½in) cutter with a 12mm (½in) shank. Choose one with 50mm (2in) cutting edges. As resin-bonded chipboards are generally used for laminated worktops, use only TCT cutters, as HSS ones will quickly overheat and become blunt. For a medium-duty router, fit a similar cutter, preferably with a 10mm (⅜in) shank.

Joining worktops

Where possible, cut sections of worktop to fit wall-to-wall. If you must join sections end-to-end, support the joint with base-unit side panels or leg frames.

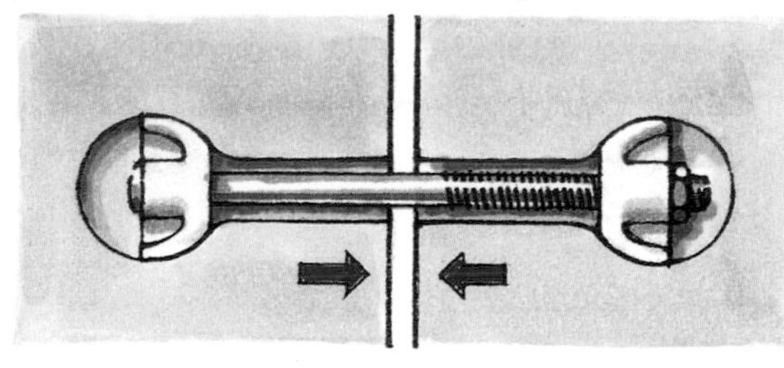

Joining sections of worktop

To ensure that the top surfaces are flush, fit biscuit dowels, using a grooving cutter to cut the dowel slots (see page 100). Apply glue sparingly to the dowel slots and face of the joint. Fit panel connectors in the undersides of both worktops, to pull the two faces of the joint together.

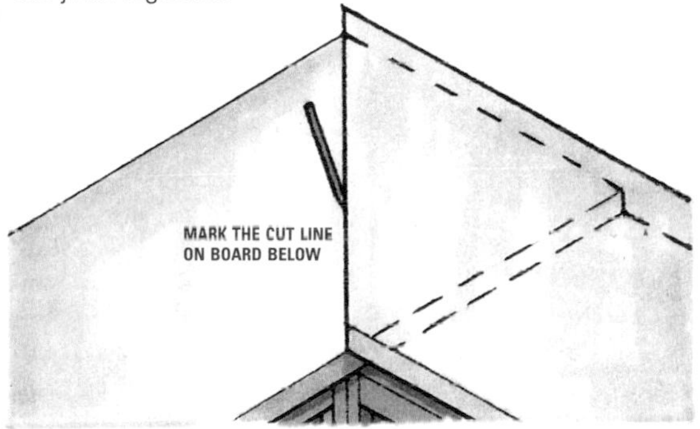

Cutting mitred corner joints

Since room corners rarely meet at exactly 90 degrees, bisect the corner angle to mark the line of a mitre joint on one length of worktop. Cut along this line, using a straightedge guide. Use the cut edge to mark the corresponding line across the remaining length of worktop. Prop and align both sections as accurately as possible before drawing the cut line with a pencil. Make the second cut as before.

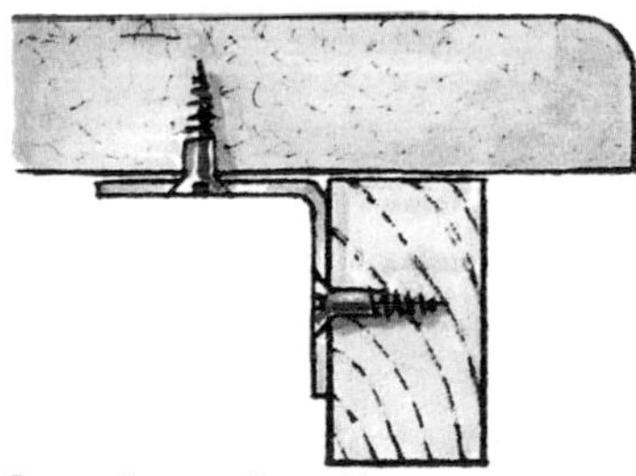

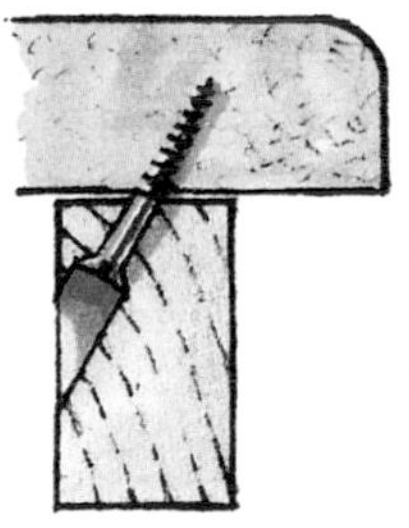

Securing and sealing

Seal all cut edges with polyurethane varnish and allow to dry. Secure worktops to the base units, using metal brackets or wood blocks, or bore screw pockets through the carcass rails. Seal joints between walls and worktop with clear silicone mastic.

MITRING BULLNOSE JOINTS

You can join worktops at a corner with simple square butt joints. However, you still need to cut small mitres to form neat corners when joining bullnose worktops.

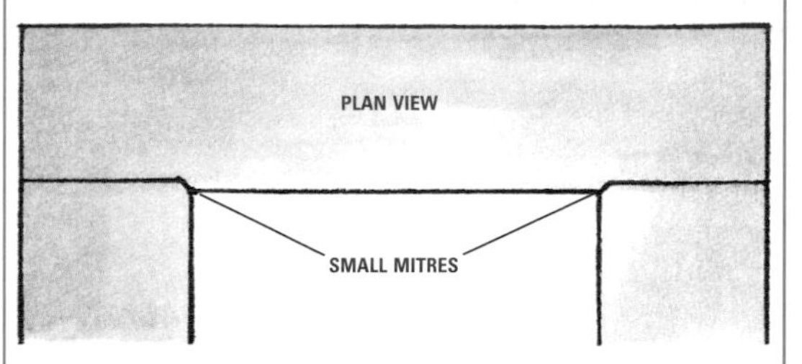

1 Mitring the front corner

Cut one length of worktop to fit wall-to-wall. Trim the end of the abutting worktop at right angles, then mitre its bullnose corner at 45 degrees, using a clamped batten to guide the router. Use a file to round over the back corner of the mitre to match the radius of the router cutter. Before starting the other half of the joint, cut this section of worktop exactly to length.

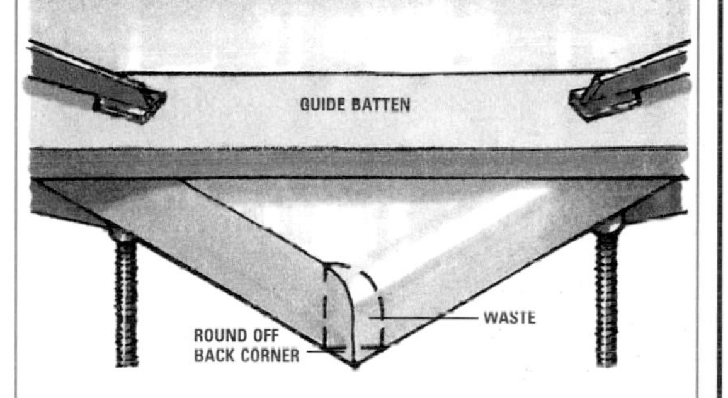

2 Cutting the matching mitre

Lay the mitred section of worktop in position, overlapping its neighbour. Use blocks to support and level the worktop, with both workpieces held tight against the walls. Draw round the cut end of the worktop to mark the other half of the joint on the workpiece below.

To cut the joint, set up a guide batten to cut parallel to the front edge of the worktop, and then reposition the batten, in order to trim the mitre.

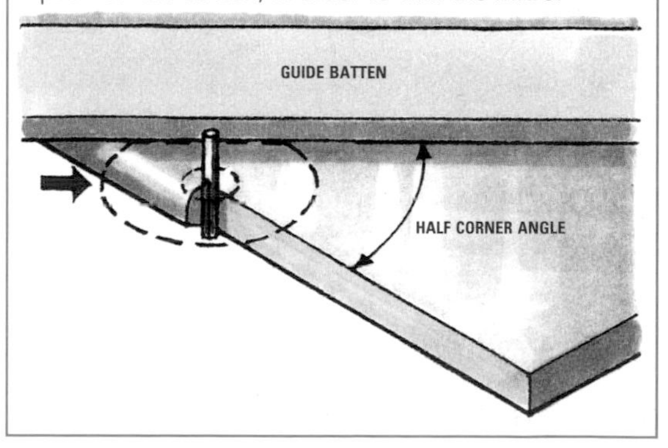

FITTING SINKS, BASINS AND HOBS

Most inset bathroom and kitchen fittings come with a full-size card or paper pattern for marking out the necessary holes on the work surface. Use a template to make safe, accurate stepped cuts.

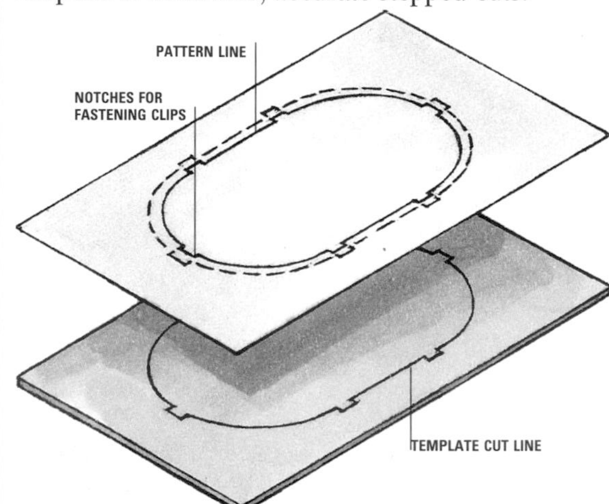

1 Making a template

The pattern gives the actual cutout size, and requires a guide-bush margin allowance (see page 50). Add this to the pattern to establish the cut line. Transfer this line onto 6mm (¼in) MDF board. Cut out the shape, using a straightedge and trammel (see pages 47 and 54). Check that the sink or basin fits the hole in the MDF, allowing for fastening clips or lugs.

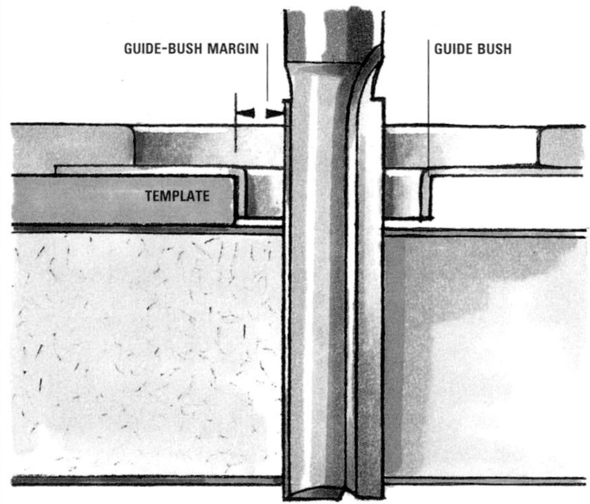

2 Cutting the hole

First check that no framing members, pipes or cables will interfere with the fitting when installed. Fix the template to the worktop, using cramps or double-sided tape. Cut through the full thickness in a series of shallow steps, working in a clockwise direction.

3 Trimming the edge

Trim the edge of the hole, using a bearing-guided profile cutter. This reduces the risk of chipping the laminate during installation, and ensures that the fitting seats securely. Seal the edges of the worktop particleboard before inserting, sealing and clamping the fitting.

LIPPING WORKTOPS

Solid-timber edgings can be applied to unfinished edges of worktops after installation.

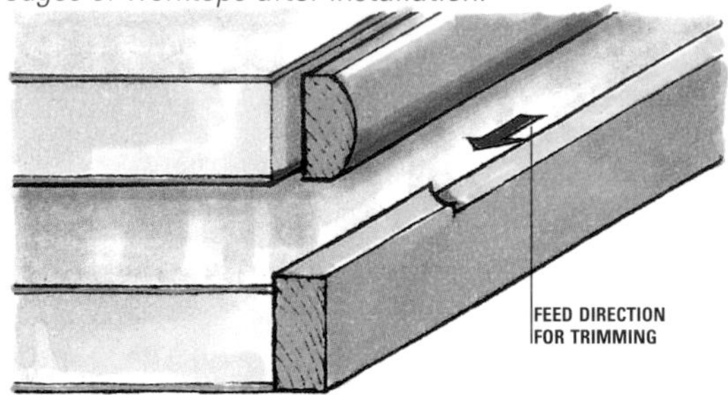

Machining the timber

Machine the timber edging to match the thickness of the worktop, and cut each section to length. Round over the edges or cut a moulding. Mark out and cut mitres at the intersections of worktop sections.

Alternatively, make the lipping deeper than the worktop and trim the top edge flush with the worktop surface after fitting.

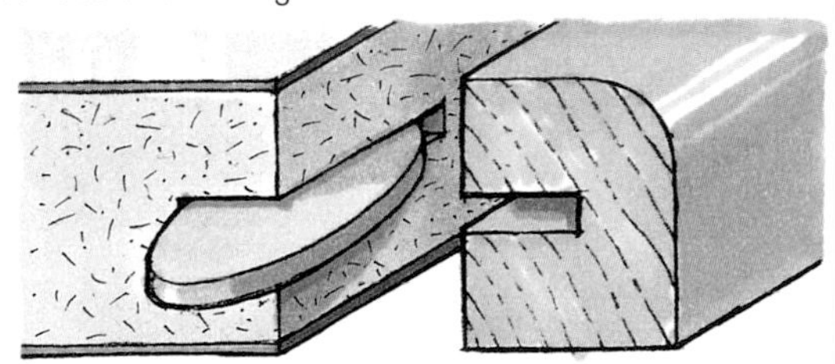

Fitting the lipping

Using a 4mm (5⁄32in) slotting cutter, set out and cut biscuit-dowel recesses at 200mm (8in) intervals along the edge of the worktop and along the back of the solid-wood lipping. Seal the chipboard edges before fitting the biscuit dowels, then glue and clamp the lipping in place.

HEALTH AND SAFETY

Regularly check your router for any signs of damage, loose parts or wear. This way, you will avoid damaging the machine and cutters, and will minimize the risk of injury to yourself and others.

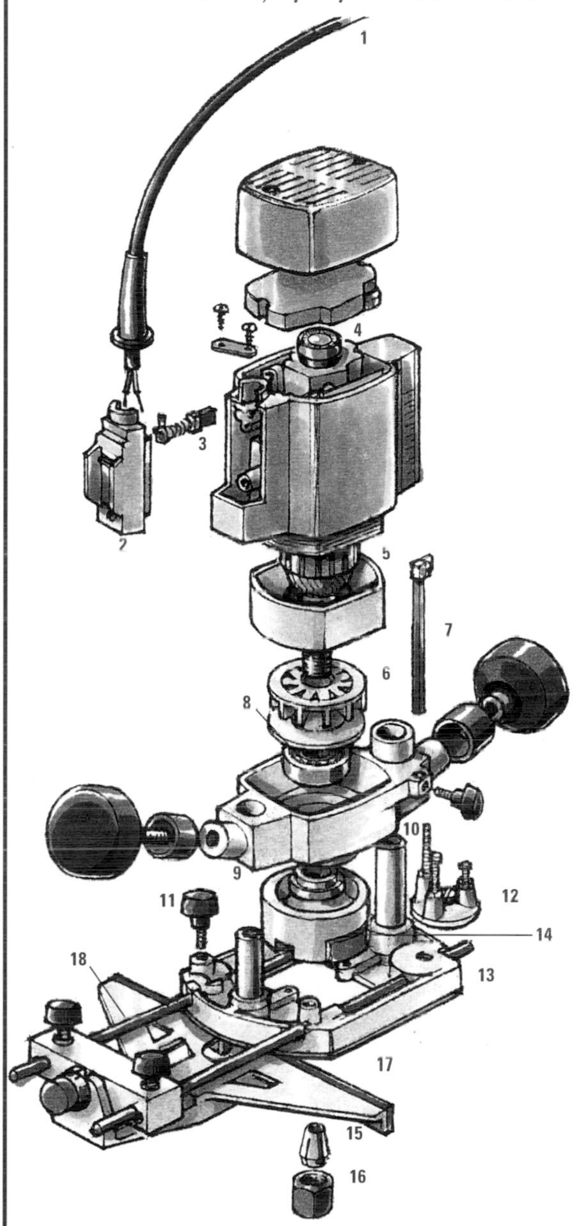

1 Mains power lead
2 Switch
3 Brush
4 Top bearing
5 Armature winding and commutator assembly
6 Cooling fan
7 Depth-stop rod
8 Bottom bearing
9 Plunge carriage
10 Plunge columns
11 Clamping screw
12 Rotating turret stop
13 Router base
14 Spindle lock
15 Collet
16 Collet locknut
17 Side fence
18 Side-fence rods

Electrical safety

Electrical accidents are often the result of poorly maintained equipment or of the operator failing to notice everyday wear and tear.

• Check that the power lead and switch are in good condition. Any sign of worn insulation, a cracked casing or the switch working intermittently or arcing (sparking), should be rectified before further use. This should be done by a qualified service engineer (particularly if the router is under warranty).

• Always replace faulty switches and cable. Never attempt to clean up burnt switch contacts, and do not wrap insulation tape around bare power-lead wires.

• Always ensure that the fuse is of the correct amperage and that any replacement power lead or extension lead is of the correct rating.

• Regularly check that the motor brushes are in good condition. Because contact between the commutator and brush springs can be dangerous, change the brushes before they wear down more than three-quarters of their length

Mechanical safety

Regularly check your router for loose or worn parts, and replace them. If you are in doubt, have your machine checked by a qualified service engineer

• Check that all clamping screws are fitted with anti-vibration springs, to prevent them working loose.

• Ensure that the plunge action operates smoothly. A cutter that jams in the plunged position or is slow to retract presents a risk to both the workpiece and the operator as the router is lifted from the work.

• Check that the collet is in good condition and that there is no undue play in the spindle bearings. Any side movement of the cutter can set up a vibration in the router, making it difficult to handle. Worn collets also lead to damaged and potentially dangerous cutters.

• Any unusual noise from bearings or brushes, or a change in the sound of the motor, usually means that something is wrong. This can be due to many reasons, such as overloading the machine through making too deep a cut, applying excessive feed pressure or using a dulled cutter. Variations in the note or pitch may also indicate a failing bearing or a cutter working loose.

Work-area precautions

Cluttered and untidy workshops increase the risk of accidents and injury. Wood shavings and chippings left on the bench or floor are not only a fire hazard, but may conceal cables and obstacles.

• Always keep the workbench and work area clear of tools and unnecessary materials, to ensure that the power cable cannot snag. Loose nails, screws and other small items may be picked up and thrown by power tools, causing damage or injury.

• Check that the workpiece is firmly and securely held, and that cramps and other obstacles do not impede the movement of the router.

• Before cutting, make sure that you can maintain your balance throughout the pass.

• Ensure that there is adequate lighting in the workshop and that you have a comfortable working temperature. Thick or cumbersome clothing can be restrictive when carrying out any wood-machining operation, and loose clothing can be dangerous.

• Don't let other people distract you. Prevent children entering the workshop and handling tools and machinery unless they are supervised.

Handling the router

Never operate a router, or any other power tool, when your are tired or unable to concentrate.

• Always unplug the router from the mains before changing cutters or fitting accessories.

• Never carry the machine by its power lead, and never unplug it by pulling on the power lead.

• Check that cutters are correctly fitted, and that all accessories and attachments are correctly assembled and secured.

• Ensure that spanners and keys have been removed from the machine. Loose tools and accessories left close to the workpiece or on table mountings and attachments may vibrate into the path of the router.

• Check that guards are correctly and securely fitted. ***In some of the illustrations in this book, the guards have been removed for clarity.***

Safe and healthy woodworking

All power-tool operations produce wood chips, sawdust, and often a high level of noise. It is therefore essential to take precautions to protect yourself against injury and damage to your health.

• Because router cutters throw out wood chips, sawdust and swarf at a high speed, always wear eye protection in the form of safety glasses, goggles or, best of all, a full-face visor. Never rely on ordinary glasses, especially if they have glass lenses.

• High-speed routers work at a level of noise that can damage your hearing. Such noise levels are also uncomfortable to work with, and can cause irritation and impatience. Always wear earplugs or muffs to reduce the noise to a comfortable level. However, it is important to be able to detect any change in the tone emitted by the router and cutter (see page 123), so ensure that ear protection does not isolate you completely from ambient sound.

• High noise levels can also be an annoyance to others. Make sure that anyone using the workshop wears ear protection while you operate a router. To avoid confrontation with neighbours or local authorities, fit efficient sound insulation in the workshop to reduce noise levels.

• Breathing in fine dust and particles can cause severe respiratory problems, even for the home woodworker who only uses routers and machine tools occasionally. Always wear a dust mask or ventilated visor. Both filter out fine particles, but a visor provides better all-round protection and is generally more comfortable.

Dust extraction

Many modern routers will accommodate a dust-extraction hood fitted close to the cutter. The hood is attached to a vacuum extractor.

In some of the illustrations in this book, the extractor hood has been removed for clarity.

• Industrial extractors can be fitted to routing tables and other stationary routing equipment, as can some domestic vacuum cleaners fitted with a fine-dust filter to protect the motor. On the best of these extractors and cleaners, an automatic feature allows the power-tool switch to operate both the tool and vacuum unit. Some also incorporate a time delay; this keeps the extractor running for a short period after the tool is switched off, to completely clear hoses and ports.